Food System Intermediaries

Food, Health, and the Environment

Series Editors: Robert Gottlieb, Henry R. Luce Professor of Urban and Environmental Policy, Occidental College

Nevin Cohen, Associate Professor, City University of New York (CUNY) Graduate School of Public Health

Keith Douglass Warner, *Agroecology in Action: Extending Alternative Agriculture Through Social Networks*

Christopher M. Bacon, V. Ernesto Méndez, Stephen R. Gliessman, David Goodman, and Jonathan A. Fox, eds., *Confronting the Coffee Crisis: Fair Trade, Sustainable Livelihoods and Ecosystems in Mexico and Central America*

Thomas A. Lyson, G. W. Stevenson, and Rick Welsh, eds., *Food and the Mid-Level Farm: Renewing an Agriculture of the Middle*

Jennifer Clapp and Doris Fuchs, eds., *Corporate Power in Global Agrifood Governance*

Robert Gottlieb and Anupama Joshi, *Food Justice*

Jill Lindsey Harrison, *Pesticide Drift and the Pursuit of Environmental Justice*

Alison Alkon and Julian Agyeman, eds., *Cultivating Food Justice: Race, Class, and Sustainability*

Abby Kinchy, *Seeds, Science, and Struggle: The Global Politics of Transgenic Crops*

Vaclav Smil and Kazuhiko Kobayashi, *Japan's Dietary Transition and Its Impacts*

Sally K. Fairfax, Louise Nelson Dyble, Greig Tor Guthey, Lauren Gwin, Monica Moore, and Jennifer Sokolove, *California Cuisine and Just Food*

Brian K. Obach, *Organic Struggle: The Movement for Sustainable Agriculture in the U.S.*

Andrew Fisher, *Big Hunger: The Unholy Alliance Between Corporate America and Anti-Hunger Groups*

Julian Agyeman, Caitlin Matthews, and Hannah Sobel, eds., *Food Trucks, Cultural Identity, and Social Justice: From Loncheras to Lobsta Love*

Sheldon Krimsky, *GMOs Decoded: A Skeptic's View of Genetically Modified Foods*

Rebecca de Souza, *Feeding the Other: Whiteness, Privilege, and Neoliberal Stigma in Food Pantries*

Bill Winders and Elizabeth Ransom, eds., *Global Meat: The Social and Environmental Consequences of the Expanding Meat Industry*

Laura-Anne Minkoff Zern, *The New American Farmer: Immigration, Race, and the Struggle for Sustainability*

Julian Agyeman and Sydney Giacalone, eds., *The Immigrant-Food Nexus: Food Systems, Immigration Policy, and Immigrant Foodways in North America*

Benjamin R. Cohen, Michael S. Kideckel, and Anna Zeide, eds., *Acquired Tastes: Stories About the Origins of Modern Food*

Karine E. Peschard, *Seed Activism: Patent Politics and Litigation in the Global South*

Jennifer Gaddis and Sarah Robert, eds., *Transforming School Food Politics Around the World*

Mariana Chilton, *The Truth About Hunger in America: And the Resolve to Heal It Through Justice, Solidarity, and Love*

Jia-Chen Fu, Michelle T. King, and Jakob A. Klein, eds., *Modern Chinese Foodways*

Adrian H. Hearn, *Food System Intermediaries: Bonding and Bridging in China, Latin America, and Australia*

Food System Intermediaries

Bonding and Bridging in China, Latin America, and Australia

Adrian H. Hearn

The MIT Press
Cambridge, Massachusetts
London, England

The MIT Press
Massachusetts Institute of Technology
77 Massachusetts Avenue, Cambridge, MA 02139
mitpress.mit.edu

The MIT Press would like to thank the anonymous peer reviewers who provided comments on drafts of this book. The generous work of academic experts is essential for establishing the authority and quality of our publications. We acknowledge with gratitude the contributions of these otherwise uncredited readers.

This book was set in Stone Serif and Stone Sans by Westchester Publishing Services. Printed and bound in the United States of America.

Library of Congress Cataloging-in-Publication Data is available.

ISBN: 978-0-262-55351-3

EU Authorised Representative: Easy Access System Europe, Mustamäe tee 50, 10621 Tallinn, Estonia | Email: gpsr.requests@easproject.com

For John Hearn and Barry Carr, mentors to many

Contents

Series Foreword

Food System Intermediaries is the twenty-fourth book in the Food, Health, and the Environment series. The series explores the global and local dimensions of food systems and the issues of access, social, environmental and food justice, and community well-being. Books in the series focus on how and where food is grown, manufactured, distributed, sold, and consumed. They address questions of power and control, social movements and organizing strategies, and the health, environmental, and social and economic factors embedded in food-system choices and outcomes. As this book demonstrates, the focus is not only on food security and well-being but also on economic, political, and cultural factors and regional, state, national, and international policy decisions. Food, Health, and the Environment books therefore provide a window into the public debates, alternative and existing discourses, and multidisciplinary perspectives that have made food systems and their connections to health and the environment critically important subjects of study and for social and policy change.

Robert Gottlieb, Occidental College

Nevin Cohen, City University of New York (CUNY)
Graduate School of Public Health

Acknowledgments

The families and activists who work the urban farms on the margins of Melbourne and Rio de Janeiro, the outskirts of São Paulo and Havana, and the fringes of Beijing are the protagonists of this book. As well as their time, they generously offered insights into the daily challenges and joys that infuse the lives of food system intermediaries. The book is inspired and informed by their work, and I am indebted to them and the institutions that generously supported its creation. Prominent among these are the Australian Research Council (ARC), the São Paulo Research Foundation (FAPESP), and the Australian Department of Foreign Affairs and Trade (DFAT). Just as critical was the advice and patience of my family as the project progressed, especially my father, John Hearn, and Adriana, Hugo, and Leo. Also instrumental was the ingenuity of friends and colleagues, especially Robin Irvine, and the freedom to write and travel furnished by the University of Melbourne. I thank each of these communities—farmers, funders, family, and friends—for entrusting me with their reflections and resources. Writing on unceded Wurundjeri country, I celebrate the wisdom of Bunjil and Birrarrung, whose traditions of land and livelihood resonate for all who listen.

Introduction: Traversing Food Systems

We cross the bridge into Manguinhos later than expected, as the distant mountains fade in the twilight. Samuel switches on the government car's internal lights, provoking memories of an earlier visit. "Here in Rio, you can't take risks," one of his predecessors had explained in 2016, "because the *tráfico* [narcotics cartel] doesn't take risks either. They need to see us so that they know who's coming into the neighborhood, and that we're with the community farm." While the protocols remain the same in 2024, the cartel has consolidated its control over this zone, diminishing the risk of conflict with competitors and the police. My involvement has also evolved, from academic researcher to supplier of horticulture equipment, like the water tanks, nets, tools, and plants packed tightly around me. Pulling up on a narrow dirt road outside the Manguinhos community association office, Samuel calls out of his window: "Finally, the special delivery: ancestral plants."

A recently graduated environmental engineer, Samuel is the latest to fill a critical yet precarious job for the city government's Hortas Cariocas (Rio Gardens) program. His main responsibility is to negotiate agreements and settle disputes about fresh food production between the Rio environment secretariat and community farmers in disadvantaged zones, where uninvited outsiders—especially police—do not enter unless seeking conflict. Two weeks into my stay in Manguinhos, negotiated by Samuel and Hortas Cariocas director Fabiola Corrêa, I have been struck by contrasting sights and sounds. Neighbors sit on doorsteps calmly chatting over a glass of cachaça, while laughing children play in the narrow alleyways late into the evening. Nobody flinches as young men gripping machine guns tear past on deafening motorbikes, while their associates sit on park benches counting towering columns of cash. Amid these urban contrasts, the community farm stretches like a green oasis beneath power transmission cables, tended to each morning by around twenty neighbors wearing Hortas Cariocas t-shirts in return for state salaries. Selecting crops, nominating work teams, collaborating with external actors (like me), and assuring the cartel that none of this will interfere with its operations falls to Samuel.

As we unload the car, Samuel records a video on his phone for his next Instagram post on ancestral plants. Since 2023, Hortas Cariocas has developed this theme to foreground Afro-Brazilian and Indigenous contributions to cuisine and pharmacology. The concept provides a measure of common ground among actors with diverging agendas, from government officials and upper-middle-class patrons of farmers markets to underprivileged food producers, foreign collaborators, and even cartel operatives. "I'll send you the video," says Samuel, aware that the Australian embassy will appreciate social media content as it funds my project over the next three years. His talent for identifying mutual interests is evident, but his employment contract will soon expire and, under current budget conditions, may not be renewed.

While Samuel is a unique individual carrying out exceptional work, he has much in common with others featured in the chapters that follow. Prominent among their skills is a capacity to articulate food traditions in ways that are precise enough to build solidarity within the groups they engage, yet sufficiently ambiguous to establish common ground between them. As the book's title suggests, their interventions strengthen inward bonding as a basis for outward bridging. This specialized work generates broadly beneficial results, and yet those who carry it out often lack professional status and job security. Embodying these contrasting characteristics, they are food system intermediaries.

Social scientists across disciplines have studied how intermediaries forge common ground. Although mediation scenarios vary in scale and timeframe, ethnographic studies show that a common undercurrent is the strategic articulation of cultural heritage to build alliances. In Bolivia's western Los Lipez region, Andrew Ofstehage observes intermediaries engaging quinoa farmers to "create a denomination of origin in order to project their work and connection with quinoa into international markets for their crop" (2011, 103). Similarly, nonprofit farmers in the United States work with "mission-driven intermediaries" to engage like-minded corporate distributors, whose "hidden reservoirs of past associations, routines, and histories" provide common ground for cooperation (Tewari et al. 2018, 122). The capacity of agricultural intermediaries to build consensus across divides is underscored by studies of international fair-trade networks (e.g., Jaffee 2007; Lyon 2011) and is conceptually prominent in recent research into the "agriculture of the middle."

Emerging attention to "the middle" focuses on the nexus of small farmers and corporate structures, particularly in the United States, as the ecological and health credentials of food systems come under growing public scrutiny (de Master 2018; Lyson et al. 2008). These intensifying interactions are provoking an "increasing demand for 'values-based' intermediaries" who broker not only the passage of food among producers, distribution hubs, and consumers but also its movement between commitments to social and environmental ethics (Thompson and Stinnett 2018, 78). Studies of resulting "value-based supply chains" show that the confluence of "right farming" and "right eating" requires trust premised on complementary priorities and shared traditions (Brislen 2018, 110; Stevenson and Pirog 2008, Campbell and Macrae 2013). Intermediaries articulate these priorities and traditions not only as messengers but also as creative agents because, as Sarah Osterhoudt et al. write, "meaning and values are not only 'added' to commodities as they move from production to consumption, but may also be obscured, transformed or erased" (2020, 2). Such proactive transformations of meaning are ubiquitous in the globalization of food as Himalayan rice merchants "become organic" (Galvin 2021), as Zambian exporters strategically package the "goodness" of "baby veg" for UK markets (Freidberg 2004), and as the intermediaries discussed in this book refashion food traditions to resonate across interest groups.

Striving in their own ways to consolidate an "agriculture of the middle," this book's protagonists depart—terminologically at least—from the classical formulations of Bruno Latour (2015, 39) and other actor-network theorists, who distinguish intermediaries (as passive "translators") from mediators (as active "transformers"). Developments in translation theory since the 1980s complicate this distinction, rejecting the notion that translators should aspire to achieve equivalent meaning (Hermans 1985). Translators, and, in my view, other intermediaries, are better understood in the sense argued by Anthony Pym (2012) as "authors" whose choices navigate the ethical risks of cross-cultural communication.

From the quinoa plantations of Bolivia to the food hubs of the US "middle," ethnographers have painted vivid portraits of intermediaries proactively transgressing boundaries to forge consensus across societies, cultures, and scales. Their studies share sensitivity to intermediaries' articulations of broadly familiar values to establish common ground among actors who may otherwise have little in common. Less explicit in these works is that intermediaries, as connective agents, function as junctures of social network formation. This is not to deny anthropology's longstanding attention to networks, which dates back to the painstaking taxonomies of kinship, ritual, and economic exchange mapped by Morgan (1871), Malinowski (1922), Radcliffe-Brown (1922), and others. Studying networks helped anthropology to move beyond the analytic dichotomy of individuals versus societies and to appreciate the give-and-take of structure and agency (Bax 1978; Sanjek 1974). Ethnographers have since shown how networks condition lived experiences of belonging and perceptions of insider knowledge, whether among Amazonian manioc "households" or amid online research groups (Collins and Durington 2015, Kawa et al. 2013). However, analyzing the structural dynamics of group membership and exclusion has become the specialty of network theorists.

Observing intermediaries helps to resolve a core puzzle in network theory: How do social groups reconcile the tension between inward solidarity and outward ties to construct networks? Robert Putnam (2000, 22–23) captures this tension by distinguishing inward "bonding" from outward "bridging," a dichotomy since employed by scholars to analyze everything from Latino ethnic entrepreneurship to shared-cost water services and socially diverse kindergartens (Berardo 2014; Smith and Mannon 2020; Vannebo and Ljunggren 2021). According to this binary model, the density of an individual's connections within their social group inversely correlates with their capacity to broker intergroup relationships. That is, people with strong allegiances to a particular community may have friends in common, but internal demands for loyalty generate "negative social capital" that impedes their ability to meet and introduce new contacts (Portes 1998, 15).

Recognizing the constraints arising from loyalty, early network theorists reasoned that people on the margins of a given group are well placed to connect its members with external partners and opportunities. Marginal social standing thus enables certain individuals to move more easily, albeit it at arm's length, among disparate groups and fill "structural holes" between them (Burt 1992). Described in the seminal work of Mark Granovetter (1973) as "the strength of weak ties," this dexterity enables such people to function as conduits for the exchange of resources through community "synergies" and institutional "linkages" (Woolcock 1998, 168). Despite the mechanical ring of these terms, their focus on intermediaries foreshadowed growing sensitivity to the cultural dynamics of network formation. As Thomas Schweizer writes, "Whereas in the past social network analysis focused on social relations, recently it has also considered social cognition or 'cultural meaning'" (1997, 740). Nan Lin, for instance, draws inspiration from anthropology's attention to "the relational aspect of exchanges" (2001, 39–40), while Laurin Weissinger argues that network scholars should consider "potentially using qualitative methods, before designing and administering a network study" (2020). The heuristic shift is evident in Neha Gondal's (2022) research into "the cultural meanings of interpersonal ties" and Putnam's (2007) conclusion that ethnic diversity influences the dynamics of bonding and bridging.

Putnam's observation of interethnic affinities, for instance, those forged by demographically homogeneous religious institutions engaged in immigrant outreach services, provoked him to refine his earlier distinction: "bonding social capital can thus be a prelude to bridging social capital, rather than precluding it" (2007, 165). This seldom-cited point raises provocative questions for network sociology and cultural anthropology, among them: How do identities and traditions intersect with bonding and bridging, how and why do intermediaries mobilize inward affinities to forge outward cooperation, how does this process shape the character of resulting networks, and how do structural conditions constrain or empower intermediaries' personal agendas?

My previous ethnographies of local government interactions with Afro-Cuban associations and with Chinese diaspora entrepreneurs in Havana and Tijuana began to explore these questions (Hearn 2008b, 2016). Those studies found intermediaries building "spheres of trust" from horizontal community solidarity and vertical policy directives, but they did not examine how distinct articulations of tradition give rise to different network structures and policy outcomes. This book picks up where those left off, showing that food system intermediaries harness the solidarity of inward bonds as a prelude to building outward bridges. They achieve this by first identifying locally meaningful yet broadly familiar traditions—for example, urban farming in Australia and Brazil, courtyard horticulture in China, and food sovereignty in Cuba—and then amplifying these to integrate the values and sensibilities of their communities with those of external partners.

The cases in the chapters that follow demonstrate that community food initiatives, when carefully mediated, can bring together public and private actors in shared pursuits. They therefore contest the conservative argument that government intervention is incompatible with a vibrant civil society. This zero-sum argument, advanced after the Soviet collapse by Francis Fukuyama (1995), Larry Diamond (1999), and others, reflects a neoliberal ideology epitomized by George Will: "As the state waxes, other institutions wane" (quoted in Skocpol 1996, 20). Studies of social capital, writes Peter Evans (1996, 3,4), came to embody this ideological trend:

> Theorists of social capital often portrayed the state as one of the culprits in the demise of community. For many theorists of social capital, the expansion of the formal bureaucratic organization of the state "crowds out" informal networks without providing the same range of value and functions, leaving communities worse off . . . some inescapable structural logic makes zero-sum relations between the state and civil society necessary or normal.

Arguments that pit state and society against each other across an imagined "public-private divide" contrast with this book's case studies and others around the world that have explored the dynamics of cross-sectoral cooperation (e.g., Fox 1996; Hearn 2004a; Nugent 1993; Portes and Armony 2018). The chapters show farming communities seeking integration with local governments and emerging markets. CERES in Melbourne was founded with government funding and has since become a leading fresh food company, supplying private customers across Melbourne; the São Paulo government's Sampa+rural project has evolved into a digital platform linking peri-urban producers with inner-city restaurants and businesses; and the Rio mayor's office subsidizes the Hortas Cariocas program on the condition that its community farms become profitable private enterprises. Even in nominally Communist systems, markets play a key role: The New Farmer program developed by Shared Harvest in Beijing has stimulated the creation of 1,000 private small farms across China, and the expanding tourism industry in Cuba has given rise to restaurants and hotels supplied by private organic farms like the Finca Marta. Such examples substantiate Theda Skocpol's observation that associational activity, entrepreneurial initiative, and the welfare state can reinforce each other in "close symbiosis" when conditions are favorable (1996, 20). Intermediaries are critical to these junctures because they actively consolidate such conditions, first by strengthening bonds within distinct groups (local officials, community farmers, private customers, etc.), and then by establishing cooperative bridges between them.

Economic arguments by themselves are unlikely to achieve such convergences, but when accompanied by narratives that resonate across cultures—for instance, traditions of nutritional health, sustainable farming, and food sovereignty—the possibilities for synergy increase. Molly Anderson (2024, 36,44) observes the centrality of narratives to food system change:

> Powerful narratives drive support for particular beliefs about food and approaches to food system transformation. One of the strongest has been the Green Revolution narrative, which claims in brief that "improved" seeds, synthetic fertilizers, pesticides and irrigation which were made available from the 1940s to 1970 in India, Mexico and the Philippines were responsible for preventing the deaths of millions of people and similar technologies could have equally beneficial effects in Africa.

Anderson shows how duplicitous narratives, promoted by institutional intermediaries like the World Trade Organization, the World Bank, and the International Monetary Fund have influenced mainstream beliefs about food production. Celebratory accounts of commodity monocrop productivity, she finds, distract public attention away from the continuing dominance of agribusiness corporations while deepening rural inequality, loss of biodiversity, reliance on agrochemicals, and malnutrition. Alternative narratives, often formulated by community associations and grassroots intermediaries, recognize these effects and seek to counteract them through progressive concepts like food sovereignty, agroecology, permaculture, and the revival of precolonial food traditions. These counternarratives reimagine the past to build consensus about the future, substantiating Anderson's observation that "the choice of how food systems should change depends considerably on how people value food and the meanings they ascribe to it" (2024, 44).

Deployed by intermediaries, narratives about food traditions function as inherently flexible signifiers. As Roland Barthes (1957) observed of a *Paris Match* magazine cover depicting a young man of African descent serving in the French military, signifiers are like leaking vessels constantly refilled with the fresh emotive content of current affairs. Politicians, business leaders, community activists, and the intermediaries who bring them together embrace opportunities to provide such content, evoking "cultural schemas" including flags, national crops, cuisines, and other signifiers to legitimize policy proposals and development projects (Hobsbawm and Ranger 1992; Ortner 1991). Food studies demonstrate how fresh emotive content constantly renews and revives established traditions, from the elevation of the disdained *gibnut* jungle rat into "real Belizean food" to the evolving ingredients of a "decent meal" in Cuba (Garth 2020, Wilk 1999). Examples like these demonstrate a blend of conscious manipulation by those who articulate traditions and unconscious collective sensibilities among their audiences. As Marshall Sahlins writes, "an invention of tradition entails some tradition" (1999, 417). Food reflects this dialectic: Its portrayal as a historical tradition harnesses public sensibilities, just as public sensibilities guide its historical portrayal.

The intermediaries who share their frustrations and achievements in the chapters that follow reveal how food traditions cut across communities and disciplines. Whereas anthropology is attuned to their use of cultural tropes to build consensus, network theory charts their role in structuring new collectives from constituent groups. A hybrid approach shows the hand of tradition in network formation as intermediaries galvanize and then harness inward bonds to forge outward bridges. A comprehensive understanding of intermediaries must grasp how these emotive and structural dynamics combine in specific cases.

Examples from Melbourne, São Paulo, Beijing, Rio de Janeiro, and Havana demonstrate that the daily endeavors of food system intermediaries are transformational. And yet their capacity to transform depends on three provisos. The first is political support, that is, the disposition of local governments to advance progressive land, labor, and environmental policies. Such policy frameworks open avenues for communities to defend urban farms and green enclaves from real estate developers by aligning their projects with municipal funding and administrative support. Although government support can help community farmers to get their projects off the ground, Anne Siebert's South African case studies signal a common global problem: "external development interventions, for instance by the municipality in promoting urban agriculture to improve food security or income creation, have a tendency to be rather short term and to ignore wider systemic inequalities, for instance precarious land or market access" (2023, 154). Cuts to public spending and associated market reforms around the world since the 1980s have exacerbated these shortcomings, prompting creative responses from the intermediaries who animate the book's case studies. Among their responses is the careful construction of narratives about urban sustainability, conservation of productive spaces, and healthy food systems that bring together public and private sources of supply and demand.

The second proviso for the effectiveness of intermediaries is collective awareness of history, namely, a measure of public knowledge and interest in the trajectory of food systems through time. A wide constellation of facts, figures, and opinions about food production circulate in the public sphere, enabling what Claude Lévi-Strauss (1962, 26–47) called *bricolage*: the repurposing of historically accrued background knowledge to make sense of current reality. He and his contemporaries borrowed the term from the avant-garde French artists of the early twentieth century, who worked as *bricoleurs* to refashion mundane everyday objects (factory machinery, kitchen utensils, textile dyes, etc.) into sculptures and paintings that offered fresh perspectives on emerging concerns. Today, in a nod to this heritage, France's leading hardware store for gardening equipment is called *Monsieur Bricolage*. The chapters that follow find that the concept aptly describes the efforts of food system intermediaries to advance collective agendas by carefully patching together historically accrued public sentiments. At their best, intermediaries are adept *bricoleurs*, cultivating the cultural undersoil to harvest sensibilities shared by those they bring together, from politicians and publics to farmers and activists.

The third proviso influencing the effectiveness of intermediaries is job security, as their work is often not officially recognized or remunerated. Some operate voluntarily alongside more conventional occupations, whereas others make a living informally in the shadows of the companies and institutions they connect. Their work has become more precarious with the proliferation of online applications designed to connect rural farmers directly with urban schools, hospitals, restaurants, and other customers. Chapter 2 will suggest that online platforms can be useful when supplemented with face-to-face interaction, but that on their own, technological solutions lack cultural sensitivity and responsiveness to unforeseen circumstances. A functional balance of digital connection and personal dialogue is possible when intermediaries are professionally acknowledged and adequately supported.

A corollary arises from the third proviso: Insecure professional status forces most intermediaries to seek personal benefits from their work. Those described in this book pursue a blend of collective outcomes and individual ambitions, including career advancement, financial gain, and personal visions of development. Self-serving ambitions invite allegations that intermediaries are free riders, whose abuses should be neutralized by cost-saving efforts to cut out the middleman. The chapters that follow illustrate some of their offences. In São Paulo's southern Parelheiros district, truck drivers (*astravesadores,* literally "crossers") notoriously inflate the cost of transporting and negotiating fresh food supplies in urban markets. To avert this risk, two Parelheiros producers—interviewed in chapter 2—bought their own trucks, but one found that their *atravesador* used it to ship deliveries for competing farms. And in Havana's increasingly competitive neighborhood markets, described in chapter 4, an intermediary entrusted with collecting and distributing fresh food misrepresents chemical-intensive crops as organic to extract higher prices. These and other examples of personal profiteering suggest that intermediaries' impacts are not always positive.

Cutting out intermediaries may diminish the potential for abuse, but it also creates new problems. As chapter 2 shows, handing over their work to marketing agencies and digital platforms further alienates consumers from the productive process and embeds profit-seeking as a legitimate and integral aspect of food systems. A more effective approach recognizes that intermediaries are justified in earning a living, and that fulfilling the third proviso—more stable and formal employment—would diminish their informal pursuit of fringe benefits.

The book explores how intermediaries balance personal and collective goals, responding to the three provisos while simultaneously influencing them. Their daily work shows that the tensions and frictions marking the interface of global agriculture and community food systems generate subtle yet powerful forms of agency. Researchers and activists are coming to recognize that this localized agency carries broad but poorly understood implications as expanding commodity plantations meet sprawling suburbs at the urban fringe. Anne Siebert (2023, 43), for instance, perceives the emerging capacity of producers and consumers in and around cities to inform international visions of food sovereignty:

> While initial steps have been taken for food sovereignty to become a galvanising concept to unite different actors, localised struggles require further attention. Deprived agrarian realities within and on the fringes of cities, including those of marginalised consumers, remain largely absent from the bigger food sovereignty picture and well-articulated discourses.

As researchers pay more attention to localized urban food systems, intermediaries will introduce them to evolving networks and invite them to engage. New questions about personal and professional ethics will arise, recalling an earlier generation of ethnographers' reflections on mediating between "the field" and the academy (Behar 1990; Clifford and Marcus 1986; Dumont 1978; Salzman 2002; Scholte 1999). Reflexivity is a valuable tool for scholars in this emerging field, especially as the social and environmental impacts of commodity export agriculture generate tensions over China's global rise. Finding middle ground between polarizing political agendas, and maintaining arm's length from their proponents, requires ethnographers to become intermediaries. How, then, do the three provisos that intermediaries accommodate and consolidate throughout the book—policy conditions, traditional sensibilities, and professional status—apply to ethnographers? Drawing inspiration from the "agriculture of the middle" theorized earlier by food system scholars, the book concludes that an "anthropology of the middle" can furnish answers.

Insights arising from intermediaries' reconciliation of personal goals and collective endeavors provide the reflexive basis for an anthropology of the middle. As with any ethnographic initiative, a useful starting point is awareness of personal bias among intermediaries, whether they work in community farms, policy think tanks, or universities. The influence of intermediaries' personal visions on their projects is evident throughout this book and, by extension, should provoke reflexivity among engaged anthropologists. The "reflexive turn" taken by ethnographers in the 1970s stirred fresh approaches, inspiring projects that aim to "bring people together in mutually beneficial ways" (Puntenney 2009, 2). In this spirit, the chapters that follow develop concepts and case studies to understand intermediaries and generate insights, elaborated in the conclusion, for an anthropology of the middle.

The Book's Chapters

Chapter 1 considers the causes and consequences of the "new commodity frontier" that is enveloping the world's food exporters. The frontier's global reach is propelled outward by what Chinese demographer Minxin Chen calls "the largest migration in human history," that is, the movement of 900 million rural people into Chinese cities over the past fifty years, and the intensification of this migration since the New-Type Urbanization Plan took effect in 2014 (Chen et al. 2021). Rural–urban migration is a cause and a consequence of the expanding commodity frontier as mechanized agribusiness—both in China and its suppliers—leaves few opportunities for pastoral employment and traditional livelihoods. Yet along the frontier's outward contours, cultural currents refashion globalization's supposedly "flattening" effects. In these transformative spaces, intermediaries broker consensus by recasting familiar traditions to make sense of unfamiliar circumstances. Urban farming is prominent among these traditions, resonating with the emerging desire of city dwellers to "reconnect" with nature while supporting the livelihoods of farmers on the urban fringe. In comparative perspective, historically accrued political sensibilities inflect understandings of China's impact, from Australian allegations of illiberal agribusiness to Brazilian charges of neocolonialism, and Cuban insistence on food sovereignty. These sensibilities serve as both backdrop and foreground for the case studies, in which intermediaries invoke China's global rise to build support for their projects from local governments, producers, and consumers.

As the contribution of urban agriculture to nutritional and civic well-being gains notoriety, real estate developers and community associations have promoted it to advance conflicting agendas. Chapter 2 analyzes this phenomenon in Melbourne and São Paulo, where colonial and industrial legacies have set the stage for a resurgence of the practice as a cherished yet contested tradition. International rankings respectively place these two cities among the world's most and least "liveable," but experiences on the ground show that despite their differences, food traditions are deployed in similar ways in both. Developers of upmarket condominiums advertise urban agriculture as a purchasable "green" commodity capable of recovering customers' lost connections with nature and each other. However, the gentrifying effects of these developments deepen rather than alleviate social and environmental ills. By contrast, community projects profiled in four case studies emphasize urban agriculture's efficacy in confronting the long-term neglect of land, employment, and environment. The chapter finds that food system intermediaries who develop narratives informed by local history attract support from consumers and municipal governments, whose alliance can stimulate policy reforms and limit corporate power. The director of CERES Fair Food in Melbourne, Chris Ennis, describes his online organics delivery platform as "a tool for public education about the social and environmental history of Australia's food system." An educational approach distinguishes Ennis and other food system intermediaries in the two cities from profit-seeking property developers, as well as from proliferating online food services. The latter efficiently connect producers with markets, but they cannot provide meaningful guidance and creative solutions—hallmarks of the *jeitinho brasileiro*—when problems arise.

Chapter 3 probes the nexus of policy, education, and profession that emerges from the efforts of food system intermediaries in Beijing and Rio de Janeiro to forge cooperation between small farmers and municipal officials. Each demonstrates what can be achieved when farming traditions handed down over generations combine with economic modeling and political acumen. Navigating insecure land tenure, outmaneuvering political opponents, fending off real estate developers, and managing food shortages deepened by COVID are daily concerns for both intermediaries and the communities they support. On Beijing's northeastern fringe, where small farms are being converted into new residential suburbs, Shi Yan strives to build an alternative food community. Faltering trust in the safety and nutrition of commercially produced food has provoked citizens to explore organic alternatives and prompted the city government to grant land and equipment for Shi's project. But to sustain this support—and her job—Yan must balance public criticism of the prevailing food system with deference to government priorities and the needs of rural migrants. Contrasting agendas run even deeper in the central Rio suburb of Manguinhos, where economic desperation among 36,000 residents generates uneasy conditions for Brazil's largest urban farm. Amid police "pacification" raids to wrest control from narcotics traffickers, Julio Barros and Fabiola Corrêa try to raise public awareness about the farm among allies and customers in wealthier districts. But when Barros's efforts attract attention from politically connected competitors allied with Rio's conservative mayor, he finds his job on the line. The political dynamics shifted with the 2023 reelection of Luiz Inacio "Lula" da Silva, enabling Corrêa to emerge as a new intermediary focused on racial and gender equity in food production. "One of the things that brings people together," she says, "is recognizing the traditional knowledge of Afro-descendants, for example through *plantas ancestrais* [ancestral plants]." But the challenges Corrêa faces, like those endured by Barros and by Yan in Beijing, demonstrate why intermediaries not only need better job security, but also less individualistic leadership structures for maintaining collaborations should they get sidelined, sick, or fired.

Sensitivity to government priorities is especially critical for intermediaries in Cuba, where centuries of extreme agricultural dependency have provoked a fierce defense of food sovereignty. Chapter 4 traces this phenomenon to

Cuba's complex relations with global powers, underpinned since the colonial era by the relentless production of sugar. US intervention at the beginning of the twentieth century entrenched Cuba's reliance on sugar exports, which was deepened by Fidel Castro's subsequent alliance with the Soviet Union. The isolation and austerity brought on by the Soviet collapse in 1989 exposed the depth of Cuba's dependency and made food sovereignty a political priority, officially articulated as a push for national self-determination. This vision underpinned Cuba's painful but necessary transition in the 1990s from large-scale commodity producer to small-scale horticulture innovator. Since then, articulations of food sovereignty have evolved to meet (and encourage) a rising tide of foreign "study tours" and to accommodate growing Chinese investments in the export sugar industry alongside attempts to grow rice and corn for domestic consumption. Food sovereignty continues to accrue new meanings that reflect changing circumstances, including assertions by private farmers about the need for faster economic reforms. These varied interpretations of food sovereignty appear in two case studies, in which two intermediaries—Luis and Reinaldo—temporalize the concept to maximize its emotive appeal and its structuring capacity. Both invoke food sovereignty to strengthen community bonds as a prelude to building network bridges, not least across geopolitical divides, for as Reinaldo points out, "Critical thinkers in Cuba and the United States want to connect with each other, and many misunderstandings can be cleared up when people see reality on the ground." The chapter concludes that history matters: How intermediaries portray Cuba's cultural past orients the purpose of their projects, the character of their networks, and the future of their food system.

The final chapter considers the role of Chinese intermediaries in communicating priorities among their government, the Chinese public, and interest groups in partner countries. Although Chinese demand is stimulating transformations of land and livelihood among its suppliers, it is less clear if resulting concerns register in the plans of Chinese enterprises and ministries. My observations and discussions with researchers and policy advisers in Beijing and Shanghai between 2007–2023 reveal sensitivity about allegations of Chinese "neocolonialism," and aspirations to improve on the historical track record of European and US predecessors. These aspirations have manifested in proposals to process soybeans into value-adding oil and animal feed in Brazil, and in investments in agriculture and energy that emphasize national sovereignty in Cuba. Chinese intermediaries bring focus to these concerns but consequently face ethical dilemmas and conflicts of interest because, as one says, "the more useful you get, the more careful you have to be what you write and say." Aligning their advice with existing policies and official narratives—such as the Belt and Road Initiative's connection with the widely admired ancient Silk Road—is a survival skill amid intense professional competition. Although their work is conditioned by the same provisos as Brazilian, Australian, and Cuban intermediaries, the balance they seek between political loyalty and personal agency carries powerful implications for their job security and for the food systems they study. The chapter finds that the ethical challenges navigated by Chinese intermediaries extend to foreign researchers who critically engage with their work. My attempts to avoid cooption into predetermined debates and political agendas—not always successful—provoke reflection on the dilemmas confronting academic intermediaries.

Reflecting on the case studies, the conclusion finds that the loss of fresh food farms caught between sprawling urban suburbs and encroaching commodity plantations is not inevitable. Resisting resulting pressures, and even stimulating policy shifts, is possible when projects foreground historical sensibilities of land and livelihood. Articulating narratives that strengthen inward bonds while stimulating outward bridges, food system intermediaries are critical to advancing community interests, protecting ecosystems, and defending food sovereignty. Their work carries insights for anybody seeking to identify and amplify common ground across distinct interest groups, not least anthropologists of the middle.

My Journey Across Food Systems

During my undergraduate years at the University of Wisconsin–Madison, in the mid-1990s, my academic advisor—the late ethnohistorian Dr. Neil Whitehead—offered me sage advice: "Before you study other cultures, first know your own history." Having moved around the world with my parents since the age of three, and later as an ethnographer, understanding other cultures and knowing my own history have become overlapping pursuits. Consequently, this book arises from life circumstances, including time living with family in Australia and Brazil, combined with two decades of research in these two countries, alongside other projects in Cuba and China. My journey, outlined in the following paragraphs, exposed me to these nations' diverse food systems but also to common threads generated by globalization. Prominent among these threads is China's growing demand for Australian, Brazilian, and Cuban commodities, which from the outset provided comparative context for the book's case studies.

My journey across food systems began in the 1980s, during primary school in London, when the government of Margaret Thatcher set about deregulating farming and cutting funding for agricultural science. Aiming to position the private sector as the engine of national development, Thatcher's policies reduced scrutiny of health and safety practices on farms while intensifying competitive pressures on their operators. The deregulation of safety standards and the opening of international markets provoked some farmers to cut corners, for instance, feeding sheep and cows with meat-and-bone meal containing sheep and cow. The consequences emerged in the 1986 outbreak of bovine spongiform encephalopathy (BSE), or "mad cow disease." One hundred and seventy-eight people died after consuming infected meat and others experienced psychological problems, dementia, and paralysis.

At that time my father was executive director of the Agriculture and Food Research Council, the sector's peak national scientific body. "I created a taskforce," he recalls, "to provide the Ministry of Agriculture with science-based advice on BSE. We needed a budget for this, but the Permanent Secretary told me the government's view: there was an oversupply not only of beef, milk, and cheese, but also of science" (interview, December 30, 2023). Thatcher continued her ideological crusade to deregulate and defund into the late 1980s, even as a second food crisis emerged.

As the government's BSE eradication program slaughtered 4.4 million cattle, the 1988 "salmonella-in-eggs" crisis erupted after farmers were found to be feeding ground poultry to chickens. The outbreak led the Under-Secretary of State for Health, Edwina Currie, to warn the public that "most of the egg production in this country, sadly, is now affected with salmonella" (BBC 1988). Although she later conceded that the extent of contamination was overstated, and would have been better described as "*much* of the egg production," the egg market collapsed as public fear set in. Currie recalled the crisis in a 2013 interview: "It resulted from laying stocks being fed 'protein' that turned out to be ground-up dead chickens. Similar insane feeding practices led to BSE in cattle in the 1980s and 90s" (quoted in Clarke 2013).

To escape the relentless cuts to funding for science and society, in 1989, my father migrated our family from London to Madison, Wisconsin. For me, the peri-urban green pastures, golden corn fields, and open blue skies were a striking contrast to the gray English megacity. But my recent visits to Madison reveal that, as in the UK, macroeconomic priorities are transforming agriculture: Those diverse landscapes of my teenage years have largely been replaced by sprawling soybean plantations. Soybeans have become the leading agricultural export of the United States (worth USD $34 billion in 2022), and China is the number one customer (UN-COMTRADE 2024).

When he visited us in Madison, my grandfather from Rio de Janeiro would cook for us his signature dish. We had come to know it as "grandad soup" during his stays with us in London: a blend of lentils, stock, and leftover pork cuttings. In Madison, it became a fusion of corn, pumpkin, and turkey. I now recognize his concoctions as a Brazilian *feijoada* stew, adapted to British and US ingredients while staying true to a tradition of versatility and pragmatism. His ancestors had left behind the urbanizing Glasgow suburb of Gorbals in the early 1800s as the Industrial Revolution intensified, settling in the Bahian city of Salvador. Like hundreds of others, their migration was facilitated by Britain's alliance with Portugal, which, in 1807, defied Napoleon's conquest of Portugal when the British navy escorted the Portuguese royal family and its entire court to Brazil. Following Prince Dom Pedro's 1822 Declaration of Brazilian Independence from Portugal, my ancestors spread south to São Paulo, Petrópolis, and Rio de Janeiro.

British migrants to Brazil brought with them particular visions of land, livelihood, and urban planning. By the early 1900s, these came to include the ideas of Ebenezer Howard and his Garden City Association, which translated into green zones for aesthetic appreciation and urban horticulture in the towns of Maringá in Paraná state, Goiânia in Goias, and suburbs such as Jardim América in São Paulo (Howard 1902). My relatives in Brazil identify with this heritage but also recognize that European colonization set in motion the nation's decimation of Indigenous communities and dependency on natural resource exports to Europe, the United States, and now China.

The ripples of Chinese demand for natural resources extend around the world, interacting with old farming traditions. The effects are acute in Australia, which relies on China—its leading trade partner—to purchase around 40 percent of its exports. In Melbourne, where I now live, the correlation of expanding commodity plantations, swelling suburbs, and waning small farms has prompted CERES and other progressive food hubs to forge new connections between peri-urban producers and inner-city customers. Observing the work of these social enterprises through site visits, interviews, and community consultations between 2017 and 2023 has revealed growing optimism, magnified by the nexus of farmers and customers during the COVID pandemic. And yet, as in Brazil, a prior colonial regime's attempt to extinguish First Nation connections with land and food laid the foundations of seemingly unstoppable industrial agriculture and mining. Of these extractive pursuits, the former's social and geographic consequences are more visible, publicly exposing the human and ecological impacts of twenty-first-century globalization.

Experiences of globalization filter through webs of culture and community, whose capacity to reshape and resist are clear in the peri-urban farms of Havana. I was drawn to Cuba by its simmering cultural and political tensions, and from 2000–2002, I lived in Havana and Santiago de Cuba to research the nexus of Afro-Cuban traditions and neighborhood development projects. At a time when the constraints of the US trade embargo impeded US-based scholars from long-term research on the island, I was fortunate to receive invitations from Australian universities to fund this work. The growing momentum of neighborhood arts projects, small businesses, private restaurants, and urban farms required their proponent to negotiate with central government ministries and foreign development agencies. "The rise of the local," as my Cuban colleagues called this swell of bottom-up initiative in the early 2000s, stimulated my collaboration with María Teresa Montes de Oca Choy and Ernel Mastrapa at the University of Havana. We convened fifteen researchers from Havana, Santiago de Cuba, and Camagüey to share their insights in a workshop and a book to examine emerging tensions, potentials, and intermediaries at the interface of state and society (Hearn 2008a).

By the early 2000s Chinese influences in Cuba, cherished but guarded from public view during the Soviet era, were becoming publicly celebrated. In step with the 2004 and 2008 visits of Chinese President Hu Jintao and a raft of bilateral agreements in transport, tourism, and agriculture, Havana's Chinatown became an attraction for national and foreign visitors. Underpinning this cultural renaissance were restaurants operated by Chinese ethnic associations, which revived longstanding family connections with food producers on the city's outskirts. Association leaders were eager to expand these operations with Chinese investment and trade, and as the prominent Chinese-Cuban military general Gustavo Chui told me, "There is no shortage of interest from partners in China in investing in Barrio Chino

[Chinatown] . . . this process should be enough to awaken the blood [*despertar la sangre*] of the community" (quoted in Hearn 2016, 153). Here lay a point of difference between the Chinese and Cuban governments: The former was eager to expand its economic presence on the island and encouraged investors—public and private—to explore opportunities, but the latter restricted cooperation to the industries identified during President Hu's visits, namely, those most tightly controlled by the state.

The prospects for deeper Chinese relations with Cuba and other Latin American nations were widely studied and reported in academic literature and conferences, but Chinese voices on these matters were rarely included. In 2007, I sought out these voices during a year in Beijing, where scholars at universities and research centers shared their views on the economic and political pressures shaping China's relations with the region and their own work. Those early discussions were conducted either in Spanish or English, as my study of Chinese language began in earnest at that time, supplemented by courses at the Confucius Institutes of the University of Sydney and, more recently, the University of Melbourne. Like their foreign counterparts, Chinese scholars recognized the need for deeper understanding of the cultural values and priorities driving economic exchanges but, unlike them, were expected to extrapolate policy advice from their research. Their "strategic scholarship," as one researcher described it, arose from the mandate of institutions including the Chinese Academy of Social Sciences, the China Center for Contemporary World Studies, and the National Academy for Development Strategy at Renmin University to simultaneously build international partnerships while advancing national benefit. This policy-oriented approach to academic research appealed to my interest in "making a difference," but it also raised ethical dilemmas about bias and positionality.

The challenge of balancing academic reflexivity with political interest and personal bias—navigated by researchers the world over—seemed to me both a stumbling block for cooperation and an opportunity for dialogue between Chinese scholars and their foreign counterparts. This opportunity was embraced by Jiang Shixue, Sun Hongbo, Yang Zhimin, Wang Ping, Zhang Jianhua, Cui Shoujun, Zhang Xuegang, and others who participated in a series of workshops and conferences at the University of Sydney, the University of California, Berkeley, the Inter-American Dialogue, and the University of Chicago. Following one of these, I was approached by the Open Society Foundations with an invitation to fund the travel and participation of four

Chinese scholars in the annual conferences of the Latin American Studies Association (LASA). These exchanges, supported further by the Australian Research Council (ARC), established a basis for candid yet respectful discussions while generating collaborative publications (Jiang 2011; Mao et al. 2011; Wang 2013; also see Strauss and Armony 2012). As Chinese initiatives go increasingly global, the trust forged through those early collaborations has sustained conversations that reconsider previous perspectives in the light of emerging projects. For instance, recent Chinese attempts to push beyond national-level agreements with member countries of its Belt and Road Initiative and build relationships with local governments and communities are positioning some Chinese researchers not only as policy advisers but also as intermediaries. How they manage the intensifying ethical conundrums, and how foreign researchers can conscientiously collaborate, are among the questions we discuss together in chapter 5.

As China's global rise generates competitive tensions with the United States and other so-called like-minded countries, engaging with Chinese colleagues is no longer a politically neutral activity. This became clear to me when I pitched a sustainable agriculture idea to a senior diplomat in Australia's foreign ministry. My proposal involved collaboration with the Shared Harvest farm in Beijing, where I had interviewed staff and customers—increasingly in Chinese—during a series of month-long research stays in 2015, 2016, 2017, and 2023 (see the chapter 3 case study). His reaction was blunt: "There's no appetite and no funding for anything China-related, so I'd suggest focusing your work elsewhere." I have encountered similar attitudes in the United States and (to a lesser extent) in Latin America, and its fallout extends to universities around the world. Academic staff are increasingly required to submit Foreign Interest Disclosure forms and the like, provoking several of my junior colleagues to indefinitely postpone their work in China to avoid unwanted scrutiny.

Amid deepening geopolitical tensions, researchers the world over must develop new strategies for mediating internal political pressures with external academic engagement. Sober appraisals of risks and benefits are prudent, but in my view, participating in Chinese teaching programs, policy forums, and research projects can open valuable opportunities to share concerns and convey hopes. A good example of this is the Green BRI and 2030 Agenda for Sustainable Development, produced by "co-leaders" from the Chinese Ministry of Ecology and Environment, the National Development and Reform Commission, and Boston University (CCICED 2020). Another, which I have gratefully participated in, is a series of Belt and Road Green Development Summits hosted by the National Academy for Development Strategy, which air the views of Chinese and foreign presenters—some of them openly critical—live on People's Daily Video (2020) and China Global Television Network to an estimated 9.5 million viewers. Chapter 5 finds that increasingly polarized assessments of China's global rise are making it harder to critically engage in public debates on the topic both in China and in the United States.

Whereas geopolitical conditions have provoked me to scrutinize my professional affiliations and personal values, applied research has energized these reflections with urgency and consequence. Working in Brazil exposed me to a high degree of community activism among scholars, who, like their Chinese counterparts, use their academic work to engage policymakers and communities. The University of São Paulo's Study Group on Urban Agriculture (GEAU), for instance, has produced pamphlets and guidelines for local governments and urban farmers to protect crops from soil and air contamination in zones with high traffic density (e.g., Amato-Lourenco 2018). Collaborating with them gave rise to a project funded by the São Paulo Research Foundation (FAPESP) and the University of Melbourne, which enabled a joint research team to visit and compare horticulture facilities and distribution hubs in the two cities between 2016 and 2019. I returned to São Paulo for three weeks in 2023, to independently develop chapter 2's case studies, which ethnographically examine COVID-19's impact on those farms and the efforts of intermediaries to rebuild networks.

From the outset of our interactions, my Brazilian colleagues and I formulated comparative recommendations for urban agriculture best practices, which we presented to the Rio de Janeiro mayor's Strategic Advisory Council in 2017. That policy-oriented approach also informed a position paper called *Food Systems and the Role of Local Government*, published by the NGO Sustain to advise municipal councils in metropolitan Melbourne (Rose and Hearn 2017). By this time, I had begun studying the Hortas Cariocas program through annual visits between 2015 and 2019 to its farms around Rio, particularly the largest in Manguinhos. Exposure to Brazilian customs, culture, and language accrued through family history and time on the ground facilitated this research, which matured into the 2024 collaboration described in the opening anecdote and in chapter 3. Two weeks of interviews and focus groups on the meaning and efficacy of "ancestral plants" informed the installation of a community medicinal plant garden in Manguinhos, bringing together academic research with the community activism favored by my Brazilian colleagues. My involvement in these projects, and the research that preceded and followed them, inevitably influenced their development and my perceptions of their efficacy. To account for this bias, each of the book's chapters offers diary excerpts and observations that, among other things, draw attention to my positioned perspectives and the awkward mishaps they sometimes produced.

My assumptions extended to Cuba, where I had previously pursued my doctoral and postdoctoral studies on the emergence of private businesses, ethnic associations, religious groups, and other civil society actors. Although the balance of government authority and community influence was shifting to the latter, researching chapter 4's case studies during visits of two weeks each (in 2019 and 2022) and five weeks (in 2024) revealed a surprising degree of agency exercised by individuals I had come to know as loyal officers of state institutions. A cumulative four years living on the island since 2001 instilled in me a degree of local expertise and language capacity, but it also generated unfounded assumptions. Using their jobs and connections in urban farming as a platform, my associates were now reaching out to forge private collaborations with foreign NGOs, tourism operators, and universities. Their expanding agency cast our relationships in a new light, leading to outcomes I had not thought possible, some of which are outlined in chapter 4. In 2017, we secured funding from the Australian government's Council on Australia–Latin America Relations for urban farming pioneer Caridad Cruz and two of her colleagues to visit community farms, distribution hubs, and local governments in Melbourne. Drawing on these insights, environmental historian Reinaldo Funes and I developed a project for the Australian Foreign Ministry's Direct Aid Program (DAP), which provided horticulture equipment for nutritional and medicinal plant production by the Afro-Cuban organization Ifá Iranlowo (SOM 2023). On the back of this experience, and as outlined earlier, in 2023, DAP funded my proposal to provide water tanks, pumps, hosepipes, and other infrastructure to a community farm in the Rio suburb of Manguinhos, the site of chapter 3's Brazilian case study (see Hearn 2023; SOM 2024).

These applied projects generated legitimate concerns from stakeholders: Funding agencies sought assurances that, as an academic researcher, I was qualified to undertake practical projects; my university colleagues wondered what publications would result; foreign ministries and ambassadors sought consultations to assess potential political motivations; and the communities I engaged asked if I could be relied upon for long-term support. Answering these questions stimulated reflection about fulfilling my personal and professional responsibilities while balancing distinct (and sometimes competing) interests, that is, how I might operate as an intermediary.

Working in China, Brazil, Cuba, and Australia has provided insights into the daily challenges faced by the intermediaries who appear in the chapters that follow. Their efforts to forge consensus among culturally and politically diverse groups offer valuable lessons for those seeking to understand—and perhaps personally engage with—the uneasy junctures of community horticulture and global agriculture.

1 Mediating the Urban Fringe

First published in England in 1734, the tale of Jack and the Beanstalk tells of a poor farmer, Jack, who (in later renditions) climbs a magical beanstalk to a celestial kingdom of riches. Jack brings treasures back down to his home, including a goose that lays golden eggs, and ultimately cuts down the beanstalk to kill a ferocious giant who is pursuing him. The beanstalk enables Jack to access otherwise unattainable benefits, but the associated danger forces him to sever the celestial connection. Although the story leaves aside the ethics of Jack's incursion into the giant's house to steal his belongings, its enduring lesson is that new connections bring both promise and peril. Not long after the story's publication, English subsistence farms like Jack's were decimated by the Industrial Revolution, but the folk lessons embedded within remain universally relevant and widely told.

Today, the fresh food farms that feed the world's growing cities do not produce golden eggs, but those who tend them depend on connections to treasures of a different kind. Among these treasures are government subsidies, tax relief, and land grants; public support for their contributions to sustainable food systems; and physical and digital opportunities to bring their produce to markets. In the absence of magical beanstalks, such connections are furnished by human intermediaries who understand and explain local concerns about land and livelihood to politicians and publics. As comfortable in the farm paddock as in the corporate boardroom, these rare individuals are at once valuable and vulnerable. Illness, jealousy, and professional conflicts are among the threats that can sever their connections, which, unlike Jack's beanstalk, sustain the livelihoods of entire communities. When intermediaries get cut off from their networks, as we will see in chapter 3, the communities they serve have no supply of golden eggs to sustain them.

Urban and peri-urban farmers depend on intermediaries—and sometimes become them—to navigate emerging opportunities and perils as the world's cities swell to a projected six billion people by 2050. Increasingly hemmed in by industrial commodity plantations on one side and by expanding concrete suburbs on the other, they work at the juncture of local agriculture and global agribusiness. Producing fresh food for urban markets, they embody an unfolding story that researchers have called the "agriculture of the middle" (Brislen 2018; Lyson et al. 2008; Thompson and Stinnett 2018).

A dimension of this story that is gaining long overdue attention is the contribution of urban farms to resilient and socially equitable food systems (FAO 2014), to the UN Sustainable Development Goals (UN-DESA 2018), and to environmental conservation (Ecocity Builders 2019; UEA 2019, C40 Cities 2019). Knowledge of these contributions is growing among consumers, who are coalescing in community food movements for reasons described by Alison Hope Alkon and Julian Agyeman (2011, 2):

> The purchase of local and organic food is cast as a "vote with your fork," to quote a common movement refrain. It is a vote for environmental sustainability, as local, organic producers cultivate biologically diverse polycultures and avoid the use of synthetic pesticides and fertilizers. It is also a vote for small, family-owned farms, as opposed to their large, corporate counterparts, and for creating local communities filled with rich interpersonal interactions. The food movement narrative also argues that linking producers and consumers will endow both with a sense of connection to local place, which allows eaters to better understand the social and environmental processes through which their food is produced.

Conscious consumers are aware that the expansion of cities and the growth of industrial agribusiness are fueling each other, generating an urgent need for local producers and for studies that illuminate the factors necessary for success. Among these factors are financial support, legal advocacy, farm equipment, and technical training, but the cases explored in this book show that the integration of such inputs requires encompassing food traditions. How such traditions are identified, deployed, and operationalized is central to the development of contemporary food systems. One of the book's core findings is that intermediaries are at the heart of these scenarios, intertwining local histories with global narratives to forge consensus among otherwise disconnected producers, consumers, and governments.

Intermediaries often articulate food traditions to promote progressive education and health policies in partnership with rural migrants and small farmers, but they must contend with alternative interpretations of history circulated by more powerful actors. Among the latter are ecological stewardship advertised by Australian and Brazilian real estate developers to validate exclusive "green" apartment blocks with gated food gardens (chapter 2), Confucian heritage reformulated by the Chinese state to legitimize top-down regulation of industrial agriculture (chapter 3), food sovereignty portrayed by Cuban officials as an emblem of the revolution to slow potentially destabilizing reforms (chapter 4), and trade along the ancient Silk Road depicted by the Chinese media to promote the Belt and Road Initiative (chapter 5). Such widely promoted narratives suggest that powerful institutions can act as intermediaries, and that their pursuit of private profit and state control diverges from grassroots visions like those espoused by individual intermediaries in the book's case studies. These contrasting agendas generate competing renditions of history that influence public opinion and, therefore, constitute a front line in the battle for food system sustainability. As Molly Anderson writes, "Narratives can be true or false, but false narratives can be quite compelling if they are promoted by charismatic figures or serve powerful interests" (2024, 34).

Food system intermediaries engage competing narratives in their attempts to forge consensus at the interface of government, corporate, and community interests. Mounting pressure on small and medium enterprises through the COVID pandemic intensified these narratives and the agendas they serve, inviting reflection on the ways intermediaries weave articulations of traditions into the formation of networks. The case studies that follow—from Melbourne, São Paulo, Beijing, Rio de Janeiro, and Havana—illuminate what it takes to forge cooperation between individuals and groups that are unfamiliar, and sometimes antagonistic, with each other. To grasp the cultural and structural dynamics of these scenarios requires an analysis of intermediaries that illuminates the influence of food traditions in the formation of networks.

Before visiting the cases and community movements supporting their work, it is useful to first consider the changing dynamics of agricultural production and consumption conditioning their development. As Eric Holt-Giménez (2011, 310) writes, "we need to examine not only these movements, but also the power structures that have shaped the global food system." Conditioning these structures in the twenty-first century is the unprecedented growth of demand for food generated by urbanization in China.

The New Commodity Frontier

Global production of agricultural commodities correlates with the demand generated by China's New-Type Urbanization Plan, whose attempt to increase the nation's urban population to one billion (from just seventeen million in 1978) is stimulating the largest migration in human history (Chen et al. 2021). The unprecedented demographic transformation has unleashed urban middle-class consumption of pork sustained by imported soy-based animal feed, as well as wheat, barley, and other commodities. For the nations that produce these commodities, balancing exports to foreign markets with sustainable local food systems raises practical and ethical challenges.

Latin American soy cultivation is concentrated in the Southern Cone, where it now accounts for forty-five million hectares, 90 percent of which are in Brazil and Argentina. As Niels Søndergaard et al. (2023, 40) observe:

> Since the turn of the millennium, Brazilian and Chinese food systems have become interlocked in a protein complex of staggering economic significance. Brazilian vegetable and animal protein has thus either directly or indirectly become a cornerstone of Chinese diets. . . . Chinese demand has become a substantial driver of socio-environmental transgressions in Brazilian soy frontiers and in many livestock-producing regions in both the Brazilian *Cerrado* and Amazon. This has not only fueled deforestation but also led to biodiversity loss and harm to vulnerable populations.

By 2010, soybeans genetically modified for resistance to the herbicide glyphosate accounted for an average of 85 percent of the total produced in the region. This has permitted intensive application of this and other agrochemicals, despite their diffusion into water systems that sustain local ecosystems and communities. "A clear outcome," concludes a multilateral task force, "is the externalization of the ecological, social and public health costs deriving from soybean production" (Catacora-Vargas et al. 2012).

Soy agribusiness has encroached on peri-urban land previously used for fresh food production, undermining ecological diversity and traditional livelihoods while accelerating rural–urban migration as land management becomes concentrated in fewer hands. Propelled by this process, Latin America's urban population has now reached 82 percent of the total, making it "one of the planet's most urbanized regions" (ECLAC 2021). Booming commodity exports have generated substantial macroeconomic benefits, but the ecological, social, and territorial effects are severe.

The expansion of soybean plantations across the Southern Cone represents a new commodity frontier, with transformative consequences for the Atlantic Rainforest in southern Brazil, eastern Paraguay, and northeast Argentina. Since the sixteenth century this area has been subjected to waves of intensive logging, gold and iron ore mining, cattle ranching, and sugar and coffee production. Warren Dean (1995, 40) observes the fundamental difference between these extractive pursuits and precolonial formulations of frontier:

> The region of the Atlantic Forest, then, presents two or three categories of frontier: that between the forest and the savanna . . . that of the first wave of humans . . . and that between the Tupi and their hunter-gatherer adversaries. The cast of characters was dramatically varied, but now another was to be added, and the plot itself was to be radically rewritten.

The extractivist pursuits introduced by European settlers propelled commodity frontiers outward into "uncommodified spaces" through a confluence of agricultural technology, local incentives, and global conditions (Joseph 2019, 21; le Polain de Waroux et al. 2018). While the extension of those frontiers was resolutely enforced by colonial and postcolonial regimes, recent work on cross-border South American commodity frontiers demonstrates that neither their geographic scale nor output match those of soybeans (Rocha et al. 2023; Stokes 2022).

The industrial success and territorial advance of soy is facilitated by new farming technologies, growing foreign investment, and neoliberal deregulation. These developments coalesce in "areas where the production of agricultural commodities (e.g., beef, soy, or palm oil) by large-scale farms expands over other land uses, the development of which is shaped by the greater ability of these large-scale actors to influence and capture agricultural rents" (le Polain de Waroux et al. 2018, 205). The resulting conversion of land echoes the colonial projects launched in the sixteenth century and subsequent US agribusiness operations, whose dominance was supported by the introduction of free trade regimes and what Gerardo Otero calls "the neoliberal diet" (2018). The impact of expanding soy plantations on land and livelihood resembles that of sugar, described in the 1980s by a rural midwife to anthropologist Nancy Scheper-Hughes (1992, 33) in the Brazilian state of Rio Grande do Sul:

> In these times we are living in, all you can see in the middle of the world is cane, cane, *cana de açucar*. That is our curse. The *roçados* and *sítios* [small traditional peasant holdings] are gone, and we have been chased away. Our lands have been "eaten up" by sugar. This is a *castigo* [punishment] from God.

The booming soy industry is the latest reverberation of this extractive echo, embodying shifting geopolitical currents and an associated change of what Harriet Friedmann and Philip McMichael (1989) have called "food regimes." Charting the correlation of agricultural production and consumption with the rise and fall of imperial powers, analyses of food regimes identify phases of agricultural governance that have served the interests of European powers (particularly Great Britain during the late nineteenth to early twentieth centuries), the United States in the mid-twentieth century, and multinational corporations since the 1970s. Recent research carries this analysis into the twenty-first century, marked by the unique dynamics of China's global rise. China is "reordering the food regime," writes Valdemar João Wesz Jr. et al., through "integration into the capitalist world system and the pursuit of a self-reliant food security policy" (2021, 1376). What distinguishes the emerging food regime from its corporate predecessor is the prominence of state institutions, ranging from enterprises and banks to think tanks and universities. As Scott Lin (2022, 290) argues:

> The hypothesis in food regime theory that the state apparatus has become subject to the agribusiness companies fails to account for the fact that the Chinese state apparatus is vigorously protecting China's agricultural markets and supporting the rise of its agribusiness SOEs.

Now spearheaded by Chinese firms and finance, the cumulative consequences of successive commodity frontiers are evident in Brazil's two largest cities, São Paulo and Rio de Janeiro, which have absorbed entire communities displaced by agribusiness operations. Harnessing their agricultural skills in urban and peri-urban farms has become a core activity for the intermediaries discussed in chapters 2 and 3, who envision mutual benefits for rural migrants, local governments, and millions of city dwellers committed to more ecologically sustainable, socially inclusive, and personally healthy food systems.

The new commodity frontier extends beyond South America to other food exporters, generating comparable challenges in its wake. The correlation of increasing exports with expanding plantations, the collapse of small farms, the intensifying application of synthetic pesticides and herbicides, and the growth of cities to accommodate displaced rural communities has become a global problem. In chapter 2, I compare how intermediaries are raising awareness about this problem and articulating food traditions to resist its impacts in São Paulo and Melbourne. Comparing these cities is instructive because their reserves of urban and peri-urban green space contrast as sharply as their so-called liveability (The Economist 2017, 2023). Moreover, Brazil and Australia harbor similar legacies of colonial land use and current dependency on grain exports to China, whose thirst for commodities has made it the main trade partner of both. Unlike other mining and agriculture-intensive nations, though, Australia has avoided the worst of the "resource curse."

Globalization Down Under

China's agricultural footprint extends to Australia, where grains have become the nation's fastest-growing export, earning USD $3.5 billion in 2020 (UN-COMTRADE 2024). Saul Eslake (2011, 145) argues that Australia is "unusual for an advanced economy" because it provides a "counterexample": Manufactured goods constitute only 16 percent of exports whereas commodities underpin economic growth, as they have since colonial times. Since the turn of the twenty-first century, demand for wheat, barley, and metals generated by China's growing cities has sustained Australia, like its South American counterparts, through successive global crises.

Australians recognize the benefits of trade with China, but they also worry about environmental and social impacts. The annual poll conducted by the Lowy Institute for International Policy (2023) canvasses public opinion on a range of issues facing the nation, providing insight into the motivations for these concerns. In 2014, 56 percent of respondents agreed with the statement that the Australian government is "allowing too much investment from China," and by 2018 the number had grown to 72 percent. Although Australians are accustomed to large Chinese investments in mining and energy, trepidation about agriculture appears to have driven the trend, with 87 percent responding in 2016 that they were against "the Australian government allowing foreign companies to buy Australian farmland." In 2021, only 6 percent of respondents—an all-time low—were in favor of investment from China. As a respondent to a *Sydney Morning Herald* survey put it, "China has polluted its waters, air and soil. They have no respect to their own motherland. Why would they care about the environment of Australia?" (quoted in Bachelard 2018). Hostility toward China has been aggravated by trade tariffs imposed on Australian barley, beef, and wine in 2020. Although farmers temporarily diverted harvests to Saudi Arabia and India, their long-term strategy remained focused on Chinese demand and was rewarded when the tariffs were rescinded in December 2023.

The 1788 arrival of the First Fleet in Botany Bay and Port Jackson (now Sydney Harbor) initiated a process of territorial dispossession that, like in South America, fractured First Nation connections with land and food to lay the foundations of industrial agriculture and mining. Of these extractive pursuits, agriculture's social and territorial dimensions are more visible, publicly exposing the human and ecological consequences of twenty-first-century globalization. Loss of crop diversity, intensification of chemical inputs, and resulting demographic and environmental impacts again reveal themselves as hallmarks of the new commodity frontier. Around Australia's fastest-growing city, Melbourne (population 5.1 million), it is striking that—as in Brazilian cities—peri-urban farms are disappearing as new suburbs extend outward and surrounding agribusiness operations encroach inward (figure 1.1). Focused squarely on commodity exports, the state and federal governments appear unconcerned that the wheat, barley, and canola plantations edging toward the city's north and west cannot fulfill local demand for fresh food. Rising land taxes leave little hope for vegetable and fruit farmers, whose capacity to supply the city is projected to fall from 41 percent of demand today to 18 percent by 2050, owing to declining availability of affordable land and the Agriculture Victoria Strategy's focus on exporting to Asia (Carey et al. 2018, 67; DEDJTR 2017).

It falls to social enterprises and nonprofit organizations to defend the viability of Australia's small farms. Among these organizations, Sustain, CERES, Cultivating Community, the Collingwood Children's Farm, the Open Food Network, 3000Acres, and Plan-It Rural have become pillars of support for Melbourne's food system. A branch of CERES called Fair Food now supplies more than a thousand Melbourne families each week with fresh fruits and vegetables sourced from around one hundred local growers, creating jobs and protecting land from real estate and agribusiness development. Fair Food's director, Chris Ennis, describes the initiative as "a tool for public education about the social and environmental history of Australia's food system" (interview September 17, 2018). As he writes in a CERES newsletter, "Over the past 12,000 years, the age of agriculture, most of us were farmers. In Australia in 1900 one in seven of us were farmers, today only one in 33 grows the food we eat" (Ennis 2012, 6).

Figure 1.1
A new suburb edges toward a farm waterhole in Melbourne's west, 2021. Photo by the author.

By locating his project in the long run of history, Ennis provides a broadly appealing narrative that has attracted support from small farmers, retail businesses, online customers, and local governments. As the intermediary at the heart of this network, his nascent alliance made its mark in Moreland City Council's AUD $34,000 (USD $25,000) *Food System Strategy*, the first government-funded framework in Australia to incentivize productive use of urban arable land (MCC 2017, 2020). How Ennis has forged consensus from competing economic, environmental, and social agendas is taken up as a case study in chapter 2. A key finding of that chapter is that articulations of food traditions that acknowledge historical injustices, including those that continue to afflict First Nation and refugee communities, constitute a powerful tool for consolidating advocacy networks.

As Fair Food and other projects gain traction at the grassroots, a surge of interest from foreign agribusiness investors has evoked impassioned counterreactions. As noted, the presence of Chinese finance and companies in the sector is generating objections not seen with mining, gas, oil, and other extractive industries (Hearn 2013, 2019). It is not surprising in this environment that the Australian federal government has blocked several significant Chinese deals in recent years as they gained public notoriety, only to approve them once media attention subsided. These include the 2015 prohibition of a ninety-nine-year lease of Darwin Port to Chinese firm Landbridge (approved later that year) and the 2015 denial of permission to sell the ten-million-hectare Kidman cattle empire to Shanghai Pengxin (approved in 2016 as a joint venture). In 2021, amid international warnings of possible environmental damage, political interference, and arrival of Chinese workers, the Australian government voided the State of Victoria's agreement to join China's Belt and Road Initiative (Chaudhuri 2018; Lloyd 2018; Losos et al. 2019). The government's view that the Initiative is "not consistent with Australia's foreign policy" reflects growing concerns about its implications around the world (Callanan 2021). Nevertheless, some Chinese and Latin American researchers, whose views are considered in chapter 5, contend that with careful management, the Belt and Road Initiative could strengthen local food systems in partner countries.

Waves of enthusiasm and trepidation about China reflect more than a decade of allegations that the emerging superpower's pursuit of food security constitutes a "land rush" among Chinese elites who "want a slice of rural Australia" (Cranston 2012; Böhme 2021; Bita 2011). Similar language has emerged in Brazil, where, according to former minister of finance Antônio Delfim Netto, "The Chinese have bought Africa and now they're trying to buy Brazil" (quoted in Powell 2011; Estadão 2010). Underpinning these allegations in both countries are simmering apprehensions about the relationship between food, land, and sovereignty. Seized on by sensationalist media and politicians, these apprehensions fuel a simple argument: China's need for food commodities has prompted agribusiness investments that threaten to covet farmland and compromise national sovereignty. An instructive contrast to this argument arises from Chinese engagement with Cuba, where colonial and postcolonial dependence on the sugar industry provoked the Cuban revolution of 1959 and disputes about the balance of dependency and sovereignty ever since. Even more than in Brazil and Australia, China looms large in Cuban attempts to find this balance.

Cuban Counterpoint

The eminent Cuban ethnologist Fernando Ortiz (1995) argued in 1940 that three centuries of national dependence on the sugar industry represented a "counterpoint" to the cultivation of tobacco. Beyond the economic contrast of sugar's stellar success as an industrial commodity compared to tobacco's limited commercial reach, he proposed a series of related cultural distinctions. These foregrounded sugar's reliance on enslaved Africans, some 780,000 of whom were forcibly transplanted to Cuba under Spain's colonial dominion, and the industry's mechanization under US control after 1898. The extractive focus of both regimes made Cuba the world's largest sugar exporter, instilling in Ortiz an acute sensitivity to power relations and their intersection with racial discrimination.

Chronicling the decline of traditional farming in the shadow of extractive agribusiness, Ortiz infused tobacco and sugar with cultural metaphors. His seminal book, *Cuban Counterpoint: Tobacco and Sugar*, depicted the domination of the former crop by the latter as analogous to the interactions of Africans and Europeans. The dynamic of power between the two is clear, but as Antonio Benítez Rojo writes, the relationship also signals a process of interaction that transcends binary opposition and outright antagonism: "It would be an error to think that T [tobacco] and S [sugar] relate to each other only in an antagonistic or exclusive way. I would say that they relate also in the complementary and diachronic sense of mutual interdependence that recalls the complexity of power relations" (1992, 173). This complexity is as pertinent to Cuba's current challenges with food and agriculture as it was during the colonial and neocolonial eras. As Cuban officials and their Chinese counterparts pursue what they call "mutually beneficial" twenty-first-century socialism, the counterpoint of dependency and sovereignty continues to shape the island's history (Hearn and Hernández 2021).

Toward the end of his life Ortiz saw his nation's economic dependence on the United States, entrenched by the sugar industry, give rise to the Cuban revolution of 1959 led by Fidel Castro. He would have also observed that the resulting exchange of Soviet oil for Cuban sugar was generating a new foreign dependency. As Frank Taylor (1989, 98) writes,

> The contradictions between plantation and nation, the economic dependency and political satellisation that had plagued pre-revolutionary Cuba had all, according to Guerra Sanchez, Ortiz and others been related to one common denominator—sugar. Yet after an initial display of sucrophobia, the revolutionary government in 1963 had adopted a strategic sucrophilia.

The changing currents of Cuban agriculture in the early 1960s reflect an embattled political system, whose National Institute for Agrarian Reform was directed by Fidel Castro's comrade Antonio Núñez Jiménez. Anticipating the revolution's later embrace of food sovereignty as a national and global pursuit, Núñez Jiménez described agricultural reform as a social and technological race against time. With classic revolutionary flair, he argued that the stakes for Cuba were comparably as high as the US-Soviet space race: "Our voyage to the moon will be to transform the nature of our country on a grand scale" (quoted in Funes Monzote 2019, 12). Despite the Cuban state's commitment to maintaining the sugar harvest, its mass mobilization of work brigades had limited impact in the absence of US demand and infrastructure investment. Sugar exports declined, but the dynamics of dependency and unrealized food sovereignty persisted.

Since the end of the Cold War in 1989, the Cuban government has pursued closer relations with China. Unlike their Soviet predecessors, Chinese strategists are determined to avoid the economic and political risks of clientelism. Therefore, although Chinese investment is helping to revitalize Cuba's sugar export industry, the deepening alliance is also building capacities—under the banner of the Belt and Road Initiative—to produce corn, rice, and other staples for local consumption. The emerging paradox of local interest and foreign influence, which chapter 4 calls "sovereign dependence," reflects a breakdown of power dichotomies reminiscent of Ortiz's counterpoint. As Fernando Coronil suggests in his introduction to the 1995 edition of *Cuban Counterpoint*: "A contrapuntal perspective, by illuminating the complex interaction between the subaltern and the dominant, should make it difficult to absorb one into the other, completing, however unwittingly, the work of domination" (1995, x1). Cuba reflects this enigma: Although it lacks economic and political influence in any conventional sense, its government continues to insist that foreign projects—Chinese and otherwise—foreground "a participatory principle of food sovereignty" (Gamboa Costa et al. 2013, 13).

The uneasy interface of state power, foreign influence, and community agency in Cuban food production extends beyond rural plantations into the urban farms that adorn Havana's street corners and courtyards. Farming for local consumption is a familiar pursuit for most Cubans, who were forced to become horticulture aficionados to survive the Soviet collapse. Inspired by peasant movements elsewhere in Latin America, a decisive national policy of "food sovereignty" provided government subsidies and land grants for a growing army of family farmers, whose ingenuity in the face of adversity earned them international respect (Altieri and Funes Monzote 2012; Premat 2009; Rosset and Benjamin 1994). Cuba's urban agriculture program was praised as an example of sustainable food production worthy of emulation, attracting offers of foreign assistance and a steady stream of visiting researchers and tourists. By the end of the 1990s, the state had authorized the creation of around two thousand domestic NGOs—with varying degrees of autonomy—to partner with foreign counterparts while monitoring their activities and limiting their influence (Cuban United Nations Association 2000, 9; Gunn 1995). Prominent among these Cuban NGOs was the Foundation for Nature and Mankind (subsequently renamed the Antonio Núñez Jiménez Foundation, FANJ), which coordinated the rollout of urban farms and to this day organizes visits for foreign researchers and allies.

The influx of foreign groups and their financial resources adds further nuance to the Cuban counterpoint. Ethnographic case studies of two Havana urban and peri-urban farms, presented in chapter 4, show that local and foreign interpretations of dependency and sovereignty vary widely and that intermediaries are stepping in to bridge the gap. To this end, food sovereignty is a valuable rhetorical resource: It is sufficiently global to resonate across different social groups while sufficiently local to allow flexibility in practice. Among the interpretations of food sovereignty that intermediaries must accommodate is a contentious but growing assertion among entrepreneurial Cuban farmers that it entails greater autonomy from the state and its multitude of regulations. Reconciling this view with more state-centric and patriotic visions, including those promoted by Chinese partners and US "solidarity" groups, is a formidable challenge.

To unify diverse actors and agendas behind a common cause, Cuban intermediaries articulate broadly familiar food traditions as they do in Australia, Brazil, and (as outlined next) China, though with a degree of political sensitivity only matched in the latter. Across each of these contexts, the need to build alliances that advocate community food systems amid changing international trade and investment relations is becoming more visible. Chinese demand for commodities is an important driver of these changes, but transformations within China are raising domestic challenges that resemble those faced by its suppliers. Far from being the monolithic command economy invoked by opportunistic politicians and media, China is home to a diverse range of actors and intermediaries who are emerging to shape their own local food systems.

Unity and Diversity in Beijing

The scenarios outlined earlier open an avenue for appreciating the mediation of bottom-up community projects with top-down political and economic power. Examining them from the ground up reveals that grassroots horticulture, contingent on local histories and agendas, can be as transformational as global agribusiness. Recognition of community agency is easily lost in debates about international trade and investment, especially when framed by politically heated accusations that China's growth is generating new dependencies among food producing nations while undermining their sovereignty.

Rejecting simplistic portrayals of China's impact in Brazil, Gustavo Oliveira (2021) writes that, "The issue is not whether Brazil is economic prey to China, but rather whether Brazilian and Chinese peasants and workers are prey to domestic and transnational corporate elites and the state actors who enable and advance their power and profits." From this perspective, the ecological and social consequences of unbridled agribusiness arise not from the actions of the Chinese state but rather from the unjust transformations of land and labor fueled by capitalist modes of production across these contexts. The implication is that urbanization is provoking challenges to local food systems within China that have much in common with those emerging in Australia, Brazil, and Cuba.

At my host family's dinner table in south-central Beijing in 2015, urbanization's impact on food was plain to see. As a de facto member of the Wang family in the suburb of Pu Huang Yu, I was required to follow a daily routine: Walk the dog with the family at 8 a.m., jam into the subway station by 9 a.m., and, most importantly, be home for dinner by 6:30 p.m. For Mr. Wang, dinnertime was a ceremonial occasion. Placing the large glass bowl in the center of the living room table, he would announce the dish he had prepared that day for his wife, daughter, and me. We enjoyed lamb, beef, and chicken almost every evening, but pork was his specialty, evident in the flare he added when exclaiming "京都排骨!" ("Kyoto pork ribs!"). I had first met Mr. Wang while living nearby in 2007, shortly after he moved his family from a corn farm in Hebei province, on the outskirts of Beijing, into the cramped high-rise apartment. Eight years on, even without the rent I was paying, his job as a clerk in the administration office of his apartment complex sustained a diet that a decade ago would have been unthinkable.

It is difficult to fully appreciate the socioeconomic importance of pork in China. The OECD (2023) calculates that Chinese pork consumption per capita increased from 23.9 kg in 2000 to 32.7 kg in 2014, declined to almost zero during COVID, and rose again to 26.9 kg by 2023. The long-term upward trend has been stimulated by "higher incomes and a shift—due to urbanisation—to food consumption changes that favour increased proteins" (OECD 2023). To diversify consumption, the government has promoted mutton and other sources of protein since 2018, but pork remains the clear favorite. As the New-Type Urbanization Plan increases China's urban population from 850 million in 2014 to one billion people by 2030, demand for pork is set to grow. To sustain the expanding herds requires soybeans processed into animal feed, forming a global chain that stretches from Mr. Wang's dinner table to South America's soybean plantations. As noted, the associated commodity frontier extends to Australia's wheat and barley fields, which provide daily carbohydrates for millions of urban dwellers like Mr. Wang, and to Cuba's sugar plantations as gleaming Chinese tractors roll in.

To Beijing's northeast, an hour's bus ride beyond the last subway station at Fengbo, a progressive community is forging an alternative peri-urban future. Shared Harvest was founded in 2012 as an independent community-supported agriculture (CSA) cooperative, and over the subsequent five years it grew to support over forty farmers on thirty-six hectares of government-awarded land. Leaving behind stagnating villages, its members have avoided the precarity of the city's construction, factory, and informal sectors. Instead, when I first visited in 2015, they were using their knowledge to produce organic pears, pumpkins, corn, sweet potatoes, okra, mushrooms, poultry, and pork for delivery to over eight hundred Beijing families each week. When I last visited in 2023, the number of farmers had doubled to eighty, and deliveries had grown to more than 1,500 families per week, stimulated by Beijing's protracted COVID lockdowns.

Shared Harvest's director, agricultural scientist Dr. Shi Yan, describes her work in chapter 3 as cultural conservation: "By providing these jobs we offer a dignified occupation that leverages the community's skills and ancient connections to land . . . we build up concentric rings of members who help small producers join the network and compete with the unfair supermarket system" (interviews, July 31, 2017 and September 12, 2023). Harnessing the capacities of displaced farmers, she is an intermediary who puts their food traditions at the center of her work. Yan presents Shared Harvest to local officials as a beacon of cultural continuity in the face of urbanization, and in return they have provided a methane biogasification plant to produce power and fertilizer and have extended her lease until 2030. The initiative's success has since inspired the creation of over a thousand community farms across China (Lyu et al. 2020). Many of these are operated by Yan's former students, and all of them are actively building public and private sector alliances to advance small-scale, nonindustrial approaches to agriculture.

Urbanization has played out differently for the Wang family and for Shared Harvest's migrant workers. Both moved from Hebei's farmlands to Beijing's sprawling suburbs, but while the former relies on food chains sustained by foreign commodities, the latter feed themselves and their urban customers with local produce. The experiences of both, explored in chapter 3, demonstrate that the need for more locally oriented food systems is as pressing in China as it is in Brazil, Australia, and Cuba.

Across the scenarios explored in this book, the deepening global challenge of food system sustainability is provoking the emergence of intermediaries to broker the shifting boundaries between expanding cities, the farms on their outskirts, and the commodity plantations that lie beyond. Exploring how they build consensus across these divides opens an avenue of inquiry that runs through the chapters that follow. In each of the case studies, the efficacy of their work is influenced by the three provisos outlined in the introduction: policy settings, public familiarity with food traditions, and professional status. Conversely, each of the intermediaries attempts to shape these provisos to favor their agendas. Chapter 2, to which we now turn, considers how they compete to this end with profit-driven real estate developers, who have become experts in advertising urban farming to sell exclusive "ecological" apartments and condominiums. In this competitive environment, the use and abuse of food traditions hinges largely on the presentation and reception of history among urban consumers.

2 Using and Abusing Food Traditions in Melbourne and São Paulo

In the gentrifying Melbourne suburb of Moreland, an eco-friendly apartment complex housing 1,800 residents has won the Excellence in Urban Renewal award. The development features rooftop solar panels, 300 bicycle racks, and a garden where the community can pick lemons and grow carrots, tomatoes, and kale. The developer's website explains that residents are encouraged to "live sustainably . . . where history meets style" (Little Projects 2015).

Overshadowed by the complex is a less noticeable 1970s brick house with a crumbling façade, worn outdoor furniture, and a front garden tended by Donata, a seventy-two-year-old Italian widow. With her sweatpants and rusting bicycle at the gate, style is not her priority, but she says it is becoming harder to resist offers to buy her house and demolish it to make way for a new apartment building. For the time being, she continues to work her thirty-square-meter garden to produce vegetables, herbs, and fruits for her family and a new "farm-to-plate" restaurant two blocks away.

Like other upmarket eateries and bars in the neighborhood, the restaurant that Donata supplies is patronized by newly arrived professionals, some of whom say they are fortunate to have secured one of the complex's fifteen available garden plots, each measuring less than one square meter. They justify inflated rents, high living costs, and limited space as the price of reconnecting with nature in an estate that promotes sustainable urban farming. Since the community garden's inauguration, though, it has been overgrown by weeds as residents complain about the absence of a water faucet, inadequate compost facilities, and lack of support from the building manager. The only thing that seems to be growing, besides weeds, is a rift between the garden's practical reality and its depiction in real estate advertising.

More than 13,000 kilometers away in eastern São Paulo, organic farmer Genivaldo is furious. Researchers at the University of São Paulo (USP) have found that vegetables sold by four of ten metropolitan sites—including his—contain unsafe levels of lead and cadmium absorbed from traffic fumes (Amato-Lourenco et al. 2016). "I've been growing organic food for forty years and my ancestors have been farmers for centuries," he exclaims through tears, "and last night the news said my vegetables are unsafe. Now my customers don't trust me!" (Interview, June 17, 2016).

The study was presented the day before in a forum on urban agriculture followed by interviews with journalists. Its findings would be relevant to other cities as vegetable gardens spring up on congested street corners and roadside patches, exposing an emotionally loaded problem that until now has been largely overlooked. Moved by the unintended consequences of their work, the researchers set about designing a manual of guidelines for clean and safe urban farming.

When I visited São Paulo in 2023, the manual had been circulated among the city's producers, and USP had established a recycling and remediation scheme with Genivaldo's wife, Dona Sebastiana, who had become the farm's manager after his death. But a new problem had emerged: The post-COVID resurgence of demand for fresh food from São Paulo schools, restaurants, and private customers had generated competition for her land. A neighbor with an expanding portfolio of farms and residential real estate had mounted a legal claim to the site, arguing that his labor-intensive approach and economy of scale would produce higher yields. He reasoned that Genivaldo's passing voided the Sebastiana's lease from the private electricity company whose transmission lines run above the property. The local farming cooperative helped Sebastiana to counter the claim by asking its members to vouch for her reputation as a defender of longstanding organic farming traditions. The strategy worked, but in 2023, the farm's future remained uncertain as the neighbor converted an adjoining building into "sustainable housing" for urban farming enthusiasts.

The previous anecdotes suggest that the unique features of food production in cities on opposite sides of the world are influenced by common experiences of urbanization. They show how the personal priorities of small farmers can contrast with the commercial ambitions of real estate developers fixated on an expanding market. Both scenarios imply ongoing battles for people like Donata and Sebastiana in the face of "ecogentrification" (Gould and Lewis 2017; McClintock 2017). Yet, the two situations also suggest that urban agriculture's appeal is gaining traction across sectors with distinct agendas, from developers and restaurateurs to scholars, journalists, visitors, and broader publics.

Researchers are finding that the diverse and often conflicting agendas that coalesce in urban agriculture illuminate the limits and potentials of progressive action. As Nathan McClintock writes, "Coming to terms with its internal contradictions can help activists, policymakers and practitioners better position urban agriculture within coordinated efforts for structural change" (2014, 147). While such efforts are shaped on the ground by unique cultural dynamics, they are also shaped by a global process of urban expansion and real estate development to house growing populations. The tension between these local and global forces has produced contradictory narratives about urban agriculture, whose core themes appear in similar ways in cities as different as Melbourne and São Paulo.

Between 2016 and 2019, a team of Brazilian and Australian researchers studied this process in the two cities for a project cofunded by the São Paulo Research Foundation (FAPESP) and the Australian Research Council (ARC). The findings of that exchange serve as a foundation for this chapter's case studies, which revisit the Melbourne and São Paulo urban farms in 2023 to reveal COVID-19's impact on producers, consumers, and the intermediaries who connect them. Together these observations show that narratives about urban agriculture are articulated differently by property developers and community activists. Its portrayal on the one hand as a purchasable personal commodity and on the other as an expression of progressive collective action hinges on the presentation and reception of history.

Australia and Brazil reflect a comparable history of food production: from indigenous subsistence farming to colonial agriculture, and subsequent reliance on transnational agribusiness to service European and more recently Chinese demand (Alejandro 1985; Hearn 2012). Both nations' historical dependence on foreign states to buy their food and mineral commodities has generated concerns about national sovereignty. The unequal benefits flowing from agriculture and mining to domestic producers and overseas consumers was a core contention in Australian and Brazilian independence movements. Similar contentions reemerged in the second half of the twentieth century as neoliberal globalization hastened the sale of government farmland to foreign investors mainly from the United States and Europe. Allegations of neocolonialism leveled more recently at Chinese investors, contested by Chinese researchers in chapter 5, reflect enduring trepidations about the undue influence of foreign states. This unresolved tension simmers below the surface of recent Australian and Brazilian calls for food sovereignty, which advocate a shift away from export-oriented industrial agribusiness toward self-sufficient community farming (e.g., Carey et al. 2022, 26; de Paula 2023; MST 2021; Rose 2013; Rose and Gaynor 2018).

Awareness of the challenges posed by industrial agriculture to national sovereignty, green space, crop diversity, and dietary health is boosting the appeal of urban farming (Amato-Lourenco et al. 2020; Siebert 2023). But gardening is not what it used to be: As the previous anecdotes show, intense competition for land, loss of subsidized council allotments and horticulture staff, and contaminated air, water, and soil are among the pressures that now confront urban producers.

According to *The Economist Intelligence Unit*, Melbourne was the "world's most liveable city" for eight consecutive years from 2011 and ranked third in 2023, while São Paulo—the Southern hemisphere's largest city—continued to grapple with lack of green space, air pollution, and entrenched inequality (*The Economist* 2017, 2023). On the ground, though, the livability criteria of a conservative magazine do not equate to sustainable urban outcomes. Economic streamlining and deregulation are generating a particular kind of livability that emphasizes marketable aspects of sustainability—including urban farming—while diverting attention away from broader structural challenges.

The phenomenon of "urban sustainability fix" identified by Aiden While et al. (2004) manifests in Melbourne and São Paulo through the proliferation of ostensibly green apartment complexes that include community vegetable gardens and micro-orchards. These features, offered at premium prices, are advertised as a means to rediscover lost connections with nature, revitalize community ties, and improve environmental sustainability (e.g., Cardim 2017; Gardiner 2017; Ideazarvos 2019; Strachan 2015). But as discussed later, lack of space and long waiting lists often result in fenced-off garden beds measuring just a few square meters, failing to alleviate social and environmental ills as promised.

By contrast, urban agriculture initiatives are emerging in both cities to more genuinely improve food systems, promoting reflection and action to deal with structural deficits. Their proponents, introduced in four case studies, attempt to mediate the interests of municipal officials, urban farming activists, and customers to rebuild the capacities and assets that have historically enabled food production in and around cities. These include the allocation and protection of productive public land, the training and employment of horticultural workers, and less polluted urban ecosystems. Promoting these agendas among actors and networks that are often unknown (and sometimes hostile) to each other, intermediaries find that their effectiveness is shaped by political and cultural provisos.

As noted in the introduction, a political proviso for intermediaries to effectively coordinate community food projects is the capacity and disposition of municipal governments to implement progressive land, labor, and environmental policies. The following cases show intermediaries engaging government support schemes by formulating narratives that resonate simultaneously with municipal priorities and smallholder interests. The traction that such narratives gain depends on a proviso of a more cultural nature: public knowledge and interest in the historical trajectory of food systems. The background knowledge that has accumulated from public sphere discourse on urban agriculture can aid intermediaries by providing raw materials from which to construct narratives—discursive *bricolage*—about community advancement. However, the same background knowledge can hinder their work when it is evoked by comparatively more powerful real estate developers to sell urban farming as a personal commodity. A fine line distinguishes the use and abuse of traditions, for, as noted in the introduction, food's historical portrayal harnesses collective sensibilities, just as collective sensibilities guide food's historical portrayal.

The chapter's first section, following this introduction, briefly considers the ethical challenges resulting from personal involvement in two of the case studies. This short reflection helps to illuminate the biases that inevitably colored the research design and the consequent limitations of my conclusions. The subsequent section finds that the growing popularity of urban farming around the world resonates with notions of traditional food production, despite the erosion of the conditions that have historically enable it. I then outline the Australian and Brazilian contexts, sketching how colonial approaches to land use in cities and their surrounds oriented the trajectory of both nations' food systems. Growing Chinese demand for commodities, arguably representing a new global food regime (chapter 1), is further inflecting these trajectories toward agricultural monocultures. The cumulative effects, evident in Melbourne and São Paulo, include rural–urban migration, the industrialization of food production and consumption, and resulting impacts on public health. I look at examples of property marketing in the two cities that promise to restore customers' lost connections with nature and well-being through private horticulture but fail to acknowledge the structural causes of these deficits. By contrast, four case studies show how urban farming enthusiasts invoke the history of their food systems to redress structural shortcomings.

The cases yield two findings, the first of which has been foreshadowed: that the capacity of urban farming to improve food systems is enhanced when intermediaries develop historically informed narratives that connect small producers and their allies with progressive municipal programs. The second is that intermediaries, for all their efficacy and value, are vulnerable actors whose effectiveness is undermined by unstable work conditions and job insecurity. Efforts to replace them with technological solutions, such as online platforms and smartphone apps, were well underway in Melbourne and São Paulo prior to the onset of COVID-19. The pandemic confirmed that professional recognition and its associated legal benefits are critical for effective intermediaries, and that the absence of this third proviso can have serious consequences. Lockdown exemptions offered to agribusiness commodity exporters did not extend to fresh food farmers and their intermediaries, whose restricted movements and interactions left entire networks of producers and consumers disconnected and dysfunctional. Together the cases suggest that stronger occupational standing, legal status, and adequate remuneration for intermediaries are logical steps for any society seeking to improve its food system.

Building the Cases

The people and projects who appear in the chapter's case studies were introduced during a collaborative attempt to discern common food system problems and solutions in Melbourne and São Paulo. Supported with grants from our funding agencies (ARC and FAPESP), the Australian and Brazilian research teams each undertook two two-week visits to small farms operating on contested lands in and around each other's cities between 2016 and 2019. Our goal was to ground what Scott Campbell (2003, 3) calls the "loose ends" of the abundant statistical data on urban agriculture in the lived experiences of urban food producers in Melbourne and São Paulo. The insights and publications produced by that collaboration provided a platform for the following case studies, which explore how the enduring efforts of key actors required them to mediate the expectations of their communities, markets, and local governments. I revisited these actors in 2023 to examine this process and ask them about COVID's impact on their attempts to renegotiate commercial channels and political support.

The first case study is from the Greater Melbourne suburb of Dandenong, where the progressive municipal government was replacing ornamental flowers in the central public park with spinach, sweet potatoes, and taro for donation to a local charity. During six visits, the initial project team elicited public responses by approaching seventeen passers-by to canvass their impressions and followed up with eight semi-structured interviews. A structured interview involving submission of preselected questions was conducted with the council's horticulture team leader, allowing us to explore government efforts to associate the initiative with Dandenong's multicultural food heritage. The outbreak of COVID as our project concluded generated unexpected results that I describe later, including an episode of a popular television show about urban farming's capacity to diminish social isolation and increase contact with nature. The pandemic also prompted an offshoot project led by my colleague Dr. Chris Williams, who orchestrated a horticulture training program with West Melbourne's United African Farm.

The second case study, from the Melbourne suburb of Moreland, examines the evolution of an online organic delivery and farmer support program called CERES Fair Food. We conducted collective discussions with groups of five to twelve Fair Food employees, most of whom were asylum seekers with insecure migration status. Described later, these discussions enabled us to survey perceptions of food production, distribution, and education as activities intertwined with social and environmental justice. Semi-structured interviews with Fair Food's director Chris Ennis and Moreland councilor (subsequently mayor) Natalie Abboud allowed us to interrogate the interface of community activism with policy development (Amato-Lourenco et al. 2020). As in Dandenong, COVID's emergence as our project concluded generated unexpected benefits for Fair Food. To keep up with booming home delivery orders, Ennis widened his network of fresh food suppliers and hired warehouse employees from twenty-six nationalities, providing jobs that he described as "their first anchor when they arrive in Australia" (interview, December 6, 2023).

The first of two São Paulo case studies explores the effectiveness of the city government's policies for procuring organic produce from peri-urban zones, including the Bloomberg-funded Connect the Dots scheme. To investigate farmers' engagement with these policies, we conducted onsite interviews with four farm managers in the southern district of Parelheiros and ten family producers in the eastern districts of Cidade Tiradentes and São Mateus. My return visit to Parelheiros in 2023, accompanied by USP food systems researcher Angélica Campos Nakamura, showed that COVID had significantly impacted producers who relied on intermediaries to connect them with urban markets.

The fourth and final case study originates in the efforts of my São Paulo project partners, independent of my collaboration, to refashion their academic research on aerial contamination of urban crops to engage policymakers and urban farmers. Their interactions with Genivaldo in 2016, described at the chapter's outset, opened an opportunity for me to follow up in 2023. Discussions with Sebastiana, her customers, and the director of the district's urban farming cooperative, AAZL, reveal the intrigues of the intervening years: the damage done by short-term media reporting of the experiment, a hostile attempt to acquire Sebastiana's land, and how her reputation as a champion of traditional agriculture underpinned her defense.

The researchers, myself included, were personally involved in the projects described in three of the four case studies, affording comprehensive access to local data and interviewees but raising challenges for ethical and reflexive conduct. My colleague Chris Williams and I had previously advised the municipal governments of Moreland and Dandenong on their food systems strategies, which in turn provided financial and technical resources for the urban farming projects outlined in the case studies. One of the two São Paulo case studies is similarly personal, describing the Brazilian team's elaboration of its earlier published work into a practical guide for urban farmers and policymakers. The ethics review processes required by our universities provoked us to consider how our personal involvement might limit our perspectives and generate preconceived conclusions. We tried to be conscious of this risk while collecting the data and writing, and as elsewhere in this book, I have included diary excerpts that foreground reflective experiences and sometimes awkward interactions.

Our positionality as researchers engaged in community projects and policy advice is evident in the research design. For instance, our decision not to interview real estate developers and instead gather data from their websites reflects a concern that conflicting interests would likely impede a genuine exchange of perspectives and, therefore, undermine the validity of resulting data. A casualty of this decision was our inability to fully explore the motivations driving developers' incorporation of horticulture gardens and orchards into their projects. The consequent epistemological risk is to limit our understanding of developers' advertising campaigns to superficial impressions of commercial motives without sufficiently exploring potential convergences of corporate objectives with public interests.

Our research and my follow-up visits found some evidence of corporate convergences with certain sectors of the public in instances of what Sarah Dooling calls "ecological gentrification" (2009). Upper-middle-class professionals and other "gentrifiers" seeking to create a "distinctive and authentic sense of place" had established common cause with developers and community organizers in pursuit of green infrastructure and new markets for fresh food (c.f. Zukin 2011, 161; Pearsall and Anguelovski 2016, 1). However, later I argue that corporate actors and community organizers differ in their visions of urban agriculture's capacity to generate public benefits and raise awareness about enabling conditions. The narratives accompanying the proliferation of horticulture in Melbourne and São Paulo demonstrate these differences. While property developers promise a quick remedy for social and ecological estrangement, the case studies show that food system intermediaries articulate the historical accrual of these problems and the need for sustained collective action to redress them.

Urban Food: A Contested Tradition

Resurgent international interest in urban agriculture results from a combination of concerns including personal health and well-being, nutrition-poor processed diets, food waste, lack of green space, and growing skepticism of industrial agriculture (Giacchè and Silva 2014; Holmgren 2018; McCarthy and Liu 2017; Nuttavuthisit and Thøgersen 2017). As rural–urban migration boosts the world's urban populations—to a projected 68 percent of humanity by 2050 according to the UN Department of Economic and Social Affairs (UN-DESA 2018)—the notion of a return to ecologically balanced, simpler lifestyles is gaining traction. The international embrace of urban agriculture reflects this trend, expanding beyond the mission statements of community associations and environmental organizations to become standard content for real estate billboards, property developer websites, and home improvement advertisements.

The marketing of fruit and vegetable gardens in exclusive condominiums exhibits a long-practiced narrative technique: Desirable conditions are advertised as lost heritage that, for a price, can be salvaged and revived. Demonstrating what Max Weber (1991 [1921], 249) called "authority from traditional custom," developers portray local food growing as having been widespread in an imagined past, but elusive today except for select property buyers. Such purposeful invocations of the past shape consumer identities, revealing agency in the passage of culture through time: A "model of tradition" becomes a "model for tradition" (Geertz 1973, 93, also see Ortner 1991). But cultural reproduction entails both memory and amnesia, for as Lyon and Colquhoun find in their analysis of food advertising:

> We require the best that yesterday had to offer, we do not require those aspects of it that were dull, tedious, uncomfortable or inconvenient . . . we are able to extract those design features, artefacts, and even experiences, that please us and forget about those that do not. (1999, 193–194)

Forgotten in contemporary urban agriculture marketing is the inconvenient truth that preindustrial food production relied on farmers markets, subsidized supply and demand, and socioeconomic bonds between cities and their rural surroundings (Steel 2009). Structural necessities included dedicated green space, trained and remunerated gardeners, and protection of soil and air in and around cities. These conditions have become increasingly treated as costs rather than market gains, failing to register in contemporary real estate advertising. The significance of this amnesia extends beyond rhetorical omission: Ignoring the conditions that were historically critical to healthy food systems deflects public attention away from inadequate land protection, socioeconomic inequality, and the contamination of urban environments. Attention is directed instead to the apparent ecological merits of developers and the personal prestige of customers. In the words of a prominent Australian real estate agent, a vegetable garden "makes a home much more emotionally appealing . . . people want a garden they can show off to their friends that says something about them" (quoted in Farrelly 2018).

While the real estate industry has become proficient in providing cities with a "sustainability fix" (While et al. 2004), urban farmers in Melbourne and São Paulo are harnessing the lessons of the past in more holistic ways. The projects discussed later simultaneously redressed structural shortcomings and educated publics about their causes. Intermediaries were central to these efforts, articulating historically informed narratives that connected community interests with municipal schemes to address the impacts of land use, employment, and ecology on urban food systems. Dealing directly with the issues that developers have avoided enabled them to integrate their projects with emerging food sustainability policies such as the Moreland Food System Strategy in Melbourne (MCC 2017, 2020) and the suite of subsidies and resources arising from the Connect the Dots initiative in São Paulo.

To contextualize the case studies, the next section considers how the development of Australian and Brazilian food systems has set the stage for urban agriculture's resurgence. Legacies of colonial land policies and more recent adaptations to Chinese demand have seen food become progressively industrialized, processed, and mass produced, with severe consequences for peri-urban family farms and nutritional health. Melbourne and São Paulo demonstrate these structural transformations, how they are omitted from commercial urban agriculture narratives, and how their detrimental impact on quality of life has begun to provoke alternative visions. Activists in both cities articulate these visions in relation to social and environmental health, distinguishing themselves from gentrifying portrayals of food traditions.

Deep Roots: Urban Agriculture Through Time

Urban agriculture responds to quantifiable problems of food security and nutritional health, but it also reflects historically accumulated relationships between people and their natural environments. Australian approaches have been influenced by British antecedents, including the rise of suburban cities with botanical gardens, pocket parks, private domestic spaces, and allotments for growing food during the Industrial Revolution (Gaynor 2006). Ebenezer Howard, an English stenographer whose ideas became globally influential in urban planning, advocated such rural–urban integration through the concept of "garden cities" (1902). His ideas were taken up in Australian towns such as Sunshine (now a suburb in Melbourne's west), where industrialist Hugh Victor McKay integrated a public park into the sprawling Sunshine combine harvester factory. Installed in 1909 explicitly to improve conditions for employees, the park "was central to his vision of a garden suburb to house a contented and respectable workforce whose life would revolve around work, church, sport, and horticulture" (Bampton 2010, 10).

The park remains to this day, now commemorated as Australia's oldest industrial garden, under the name H. V. McKay Memorial Gardens. Perhaps more significant than the garden itself, though, is the less well-known story of grassroots activism that forced its creation. Three years prior to the park's inauguration, McKay lost a court case brought by his workers, who demanded conditions commensurate with "the normal needs of the average employee regarded as a human being living in a civilized community" (Parliament of Australia 1907, 3). The ruling was the basis for Australia's minimum wage law and probably motivated McKay's experimentation with the garden city concept to begin with.

Other British influences on Australian urbanism have their roots in nineteenth-century Manchester, Birmingham, Leeds, and Liverpool. As hubs of the Industrial Revolution, these cities experienced rapid growth and the proliferation of disease-ridden slums. This provoked parliamentary inquiries into sanitation, health, fresh air, and accessibility of urban green space. Victorian era reformers were determined to bring the country into the city and, in 1887, passed the Allotments Act to promote community food plots. By 1895, there were 450,000 allotments in Britain, and during the First World War, the every-man-a-gardener campaign increased the number to 1.5 million (Crouch and Ward 1997). Local governments employed gardeners in parks, ran nurseries, and created horticultural services. Urban agriculture thus reflected not only wartime necessity but an attempt to integrate the natural beauty of the countryside with the commercial infrastructure of towns and cities as Howard had advocated. After a decline during the interwar period, the Dig for Victory campaigns of World War II saw the number of subsidized allotments return to a historic high. The campaign aimed to secure food for working-class people while keeping them occupied with a broadly beneficial pursuit (rather than drinking or gambling), though grassroots activists pushed for access to allotments independently of this discourse (Willes 2014). These experiences conditioned overseas colonial practices, setting historical precedent for Australian urban food production, and even making their mark in Brazil.

Portuguese colonization of Brazil in the sixteenth century triggered an extensive exchange of plants with Europe. Native varieties of yam, taro, cassava, peanut, sweet potato, pine nut, pumpkin, and squash were introduced into settler diets and came to characterize Brazilian cuisine. From Europe came sugar cane, lemons, oranges, mangos, coconuts, figs, pomegranates, lettuces, cabbages, turnips, carrots, cucumbers, spinach, onions, mustard, tomatoes, and ginger. This diversification produced genetic mutations that enabled foreign species to adapt to Brazil's edaphic and climatic conditions (Madeira et al. 2008). São Paulo and other colonial settlements were initially dependent on Portuguese food supply, but transportation across the Atlantic and even from surrounding rural zones was costly and impractical. Vegetable gardens and urban farms thus became important sources of food, contributing to the consolidation of autonomous Brazilian markets, services, and systems of land governance.

When Brazil declared independence from Portugal in 1822, Britain was its main commercial partner. The ensuing century saw British investment in Brazil's industrialization and infrastructure through projects such as the São Paulo Railway, completed in 1867. Less well known is the influence of Ebenezer Howard's garden city movement on the establishment of several São Paulo suburbs, including Jardim América in 1917. The suburb was designed by the British architects Raymond Unwin and Barry Parker, members of Howard's Garden City Association who had previously been instrumental in creating the world's first garden city in Letchworth, England, in 1903. Contracted by the City of São Paulo, Parker and Unwin left their mark in Jardim América's tree-lined streets, curved boulevards, and public parks, which in turn influenced the neighboring suburbs of Jardim Europa and Jardim Paulista, and others such as Alto da Lapa (IGCI 2020). Today these zones remain a green oasis amid the megacity's gray and brown high-rises, but public access has become increasingly restricted as occupants of new condominiums gain private rights to orchards and vegetable gardens. As in Melbourne, green spaces for recreation, horticulture, and the supply of fresh food within and around the city have diminished with inward densification and outward sprawl (Duran et al. 2013).

The challenges of fresh food production in Melbourne and São Paulo reflect the adoption of neoliberal policies between the 1950s and 1980s, which unleashed private property ownership across Europe, Australia, and the Americas. Suburbs with residential gardens and backyards expanded, but the intensity of localized food production declined as these private spaces ostensibly diminished the need for public allotments. By the early 1980s, municipal governments were abandoning subsidies for allotments and horticulture training programs and scaling back to the barest standards of maintenance for public parks and gardens. Meanwhile, agriculture ministries increasingly focused on boosting chemical-intensive commodity crops for export while large retail supermarkets came to dominate domestic markets with processed and packaged food (FAO 2003; Parham 2015).

A study conducted by the Victorian Eco-Innovation Lab finds that the concentrated control of agricultural production, retail, and export by a small number of corporations is generating critical vulnerabilities in and around Melbourne (Murphy et al. 2022, 32–38,64). However, this observation and associated social and ecological contentions have little traction in mainstream politics and media, while more alarmist assertions—often similar to those articulated in Brazil—are widely circulated. Reports of an "alarming level of Chinese ownership of Australia" resemble accusations that China is "attempting to buy Brazil" (Estadão de S. Paulo 2010, Parsons and Stevens 2020). Chinese investment is regularly portrayed in both countries as an attempt to grab land, enable an influx of Chinese workers, and undermine national sovereignty. This externally directed blame, writes Michaela Böhme (2021, 441), has drawn attention away from questions about Australia's dependence on "productivist" agriculture:

> Concerns over the assumed "state-driven" and "food security"-oriented nature of Chinese investment have been mobilized to tighten governmental scrutiny over foreign investment flows into Australian agriculture. . . . Ultimately, these controversies have precluded a substantial reflection upon the problematic productivist assumptions that characterize common understandings of who should own Australian farmland and to which ends.

Apprehension about state-centric Chinese agriculture investment reveals discomfort and alarm as the global food regime shifts away from corporate neoliberal norms. But as Böhme implies, the local consequences of extractive agribusiness remain unchanged and uninterrogated. Loss of crop diversity and employment, growing rural–urban migration, and increasing "food miles" are among the enduring consequences confronted by intermediaries in the following case studies.

The corporate food regime that has dominated Australian agriculture since the 1960s reveals that the withdrawal of public funding for fresh food production correlates with the expansion of the processed food industry and a rise in dietary health problems. According to the Victorian Department of Health (VicHealth 2023), 68 percent of adults living in the state are overweight or obese, while the Ministry of Health of Brazil (2016, 14) reports 54 percent for São Paulo. Type II diabetes and other diet-related diseases are growing across Australian and Brazilian cities, a problem compounded by the vertical integration of food wholesale and retail companies. As Susan Parham writes:

> In the early 21st century, we primarily rely on intensive, chemically dependent and, now increasingly, genetically modified food production, intensive processing and packaging of food, and long-distance transportation, with enormous wholesaling facilities to serve very large-scale, car-dependent industrialized (and now also "functional" and "nutrimedical") food retailing. Food companies seek vertical integration from farm to plate, as far as possible to externalize environmental and social costs. And these costs are substantial, including impoverishing producers, diminishing consumers' tastes, and creating unacceptable food miles and food deserts through profit maximizing spatial practices. (2005, 89)

As consumers around the world become more aware of these conditions and discerning about the personal and political implications of their diets, demand for locally produced organic food is growing (Hempel and Hamm 2016; McCarthy and Liu 2017; Nuttavuthisit and Thøgersen 2017). Edible landscapes and vegetable patches are appearing in private homes and alongside suburban streets, increasingly supported by municipal governments to improve local health and sustainability indicators.

Restaurants have pursued the expanding market for organic meals featuring ingredients produced locally and responsibly. Some of these are social enterprises that source inputs from—and return organic waste to—nearby community gardens, employ disadvantaged refugees or former inmates, and communicate their vision of sustainability on the backs of menus or online. Others focus narrowly on market opportunity, conspicuously displaying planter boxes to demonstrate sustainability, charging premium prices for pretentious menus, and retailing mass-produced homewares at the counter. The two varieties of restaurant articulate their purpose differently, the prior seeking to provoke reflection about the place of food in urban renewal and the latter concerning itself mainly with public image. The growing popularity of organic menus is thus underpinned by diverging priorities (Poulston and Yiu 2011).

Real estate developers in Melbourne and São Paulo are also targeting the urban agriculture boom, incorporating private vegetable plots and gardens into their premium offerings. Seemingly progressive projects claim to revive citizens' connections with land, community, and food production but fail to stimulate awareness or action around the place of food in urban development. More genuine attempts to educate citizens about the historical trajectory of food systems and involve the public in neighborhood projects are nevertheless beginning to emerge. Recent initiatives in both cities, some of which are profiled in the case studies, have demanded—and sometimes achieved—reforms within and beyond their localities by formulating historical narratives about urban agriculture that engage consumers and municipal governments.

The case studies demonstrate that building alliances among farmers, governments, researchers, and consumers requires narratives that resonate across these distinct sectors. Teasing out a shared sense of purpose to achieve what Peter Evans (1996) calls "development strategies across the public-private divide" is a matter of forging shared purpose from multiple agendas. This delicate construction of consensus is evident in Melbourne, where two intermediaries find themselves competing with real estate developers to inform governments and publics about the past and future of urban agriculture.

Making Space in Melbourne

With 85 percent of Australia's twenty-five million citizens living in sprawling suburbs, household food production is widespread. The large size of Australian front and backyards has historically tempered demand for public allotments and community gardens, which have never enjoyed comparable support or continuity as in Britain or the United States (Gaynor 2006). Britain's earlier experiences with urban industrialization nevertheless influenced Australian approaches to land use, for instance, ensuring public access to green space in the colony's emerging cities.

After a globally significant gold rush in the 1850s, Melbourne came to embody Victorian era confidence in urban development as a marker of national progress. Embracing the philosophy of large parks and gardens, the colonial government mandated the creation of urban green areas and facilities such as the Burnley horticultural college, six kilometers from the city center on the Yarra River. The physical maintenance of these properties required coordinated workforce planning, training, and employment. At the turn of the twentieth century Burnley employed over one hundred educators and groundskeepers. Commitments to the public resourcing of green areas persisted for over a century, but the economic rationalism of recent years has disrupted Melbourne's urban ecological heritage.

Australia has one of the highest rates of urbanization in the world, with more than 75 percent of the population living in five cities: Sydney, Melbourne, Brisbane, Perth, and Adelaide. Melbourne is projected to overtake Sydney as Australia's most populous city by 2028, with migration from rural areas and overseas expected to underpin population growth from five million in 2020 to eight million in 2050 (Longbottom and Knight 2018; Urban 2018). One Australian study correlates cardiometabolic health problems related to blood pressure, obesity, and diabetes with time spent driving, noting that 78 percent of adults use a car as the main form of transport to work (Sugiyama et al. 2016). While Melbourne's incorporation of seventeen new suburbs between 2018–2023 has caused longer commutes, a simultaneous contraction of green spaces within the city is compromising quality of life. The predicament is captured in the title of Tony Hall's (2010) book, *The Life and Death of the Australian Backyard.*

The challenge of balancing outward expansion with inward densification was central to the state of Victoria's 2022 elections, as both are necessary to provide housing for a growing population, ensure adequate infrastructure as the population ages, and avoid unmanageable congestion (Urban 2018). Less well publicized is the sobering projection that Melbourne's locally produced fresh food will diminish from 41 percent of demand in 2018 to 18 percent by 2050 (Carey et al. 2018, 67). Meanwhile, a survey of over 235,000 Australians reports that overconsumption of cakes, chocolate, fast food, and alcohol has earned a "healthy diet score" of 55 out of 100 (Elder 2023).

State and federal policymakers have subordinated the health benefits of localized food production to macroeconomic priorities, such as exporting wheat and barley to China, where "demand from the emerging Asian middle class presents a major opportunity for growth in Victorian agriculture" (DEDJTR 2017, 5). Registering AUD $4 billion in annual sales to China, the Victorian government's trade agency Global Victoria (2024) sees opportunity to service growing Chinese interest in plant-based meat and dairy:

> As China seeks to enhance its food security, it is diversifying its protein sources and increasingly turning to alternative proteins. As an emerging global hotspot for food innovation, Victoria is well-positioned to service the growing demand for plant-based meat and dairy alternatives, . . . Victoria is Australia's largest supplier of food and fibre products to China. Since the China-Australia Free Trade Agreement (ChAFTA) took force in 2015, significant export barriers have been removed, and Victoria's agribusiness sector is well placed to boost sales to China.

China's nascent plant-based meat and dairy industry suggests the desire of the nation's expanding middle class to reduce meat consumption. But its reliance on wheat and barley as base ingredients is likely to deepen existing challenges faced by Victoria's small farmers, including rising land taxes, exposure to agrichemicals, and aggressive takeovers by commodity planters. As the Melbourne case studies show, the declining viability of rural towns and the quickening pace of migration to Melbourne are provoking intermediaries to articulate narratives about the loss—and revival—of food traditions that build consensus among fresh food producers, consumers, and local governments. The capacity of such narratives to unite disparate actors is evident in the Victorian Eco-Innovation Lab's recommendations for protecting Melbourne's food system. Codesigned with government, industry, and civil society, they include the suggestion to "Embed Indigenous self-determination and food sovereignty at the centre of Victoria's food resilience plan" (Carey et al. 2022, 26).

Agreement across the public, private, and civil sectors that Indigenous food sovereignty should be a priority constitutes a starting point for action. Among the associated recommendations are "growing indigenous foods that are well adapted to local ecosystems" and ensuring that First Nations communities have "access to land, resources and capital to support self-determination in establishing Indigenous-led food and farming enterprises and to restore their leadership of the Victorian native foods industry" (Carey et al. 2022, 57,63). Achieving these progressive aspirations in a context of export-oriented industrial agriculture and real estate development is a slow-burn challenge, but examples of progress are beginning to emerge. Australia's Indigenous heritage and the food traditions of underserved migrant communities figure prominently in the case studies.

The effects that small projects can have on personal happiness and community welfare became evident during the COVID pandemic. Data from the Australian National Pandemic Food Survey, conducted by the nonprofit organization Sustain, show that growing food at home and in community gardens helped to alleviate financial stress and social isolation. Of 9,140 respondents, more than 80 percent indicated that urban farming made the pandemic less distressing because it "increased food security, heightened wellbeing, and improved social connections" (Phillips et al. 2021, 2, also see Dun et al. 2018). Other studies found that during the first year of COVID, 22,651 people relocated from Melbourne to rural Victoria, where "nature is the new destination" owing to availability of green space and less stringent lockdown measures (Millar et al. 2020; Wright and Duke 2021). The resumption of face-to-face work in 2022 and 2023 forced many to return, provoking a housing affordability crisis and amplifying demand for access to urban vegetable gardens, fruit micro-orchards, and green space.

As public interest in urban farming grows and subsidized green space shrinks, private real estate developers are positioning themselves as providers of an ecological solution. An article featured on the website of Australia's largest estate agent Domain explains why: "Food for thought: sustainable eating becomes property selling point across Australia" (Farrelly 2018). Property giant Stockland, for instance, has worked with the Green Building Council of Australia to develop "Green Star" projects. According to its website, "Stockland is one of the largest diversified property groups in Australia with more than [AUD] $15.8 billion of real estate assets" across the residential, commercial, and industrial sectors (Stockland 2017). Conscious that one-quarter of Australians were born overseas, it installs community gardens that purportedly enable residents to experience "how different cultures grow things differently" in "really old-style communities" (Lindsay 2016; Strachan 2015). Other developers, such as Shayher Group, proclaim that installing vegetable gardens amid their high-rise apartments will "draw locals together" and enable them to "remain connected to nature" (Gardiner 2017).

Efforts to revive old-style traditions and connections with nature sound progressive, but their subordination to commercial priorities perpetuates systemic deficiencies. Green space and the resources required to care for it are among the casualties of the emerging trend, evident in the gentrifying central Melbourne suburb of Moreland. By 2023, eight years after the completion of the eco-friendly complex described at the chapter's outset, the property's corporate office had replaced three disenchanted site managers. In my interviews with each of these in 2018, 2020, and 2023, they lamented that the sense of community envisioned and promoted by the developer had failed to emerge. Seventy percent of residents were renting—staying two years on average—and the vegetable garden once inaugurated as a hub of social interaction had been gradually abandoned.

Residents were initially enthusiastic, entering to pick lemons and bring food scraps and other organic waste from across the complex to the garden's compost bins. Their optimism generated a waiting list of two years to secure one of the fifteen available garden plots, despite their limited dimensions (one square meter). The developer and local estate agents had expertly sold the idea of sustainable urban farming, and consumers had eagerly bought it. For university student Beth, renting one of the estate's numerous one-bedroom apartments was "expensive for the space you get, but I like living somewhere peaceful and green . . . that supports the environment." Mike, a finance manager from Ireland, said he and his wife bought one the large apartments—advertised as "green-facing" because it overlooks the garden—because "it reminds us of Cork, where we used to grow food outside our house."

Just two months after the garden's opening ceremony, the site manager started receiving complaints, which he addressed by posting rules and guidelines in public areas. One of these notices prohibited carrying buckets of water in elevators, which gardeners had been doing because the developer had not installed a faucet for irrigating their plants. A letter dropped in residents' mailboxes then forbade composting because the abundance of food scraps had overwhelmed the small bins and attracted rats. To enforce this, a lock was fitted to the garden gate to restrict access to residents with a registered plot. By then, the absence of water and compost had led most to give up their leases. Lemons now accumulate and decay under the trees, and weeds invade where carrots and tomatoes previously sprung.

The fate of the community garden indicates a lack of attention and commitment to the practical realities of urban farming. Inadequate infrastructure, space, and oversight at the complex contrasts sharply with the neighboring 1970s brick house and its thirty-square-meter front garden described at the chapter's outset. Donata chats with passers-by amid vibrant green onion shoots, yellow lemons, and an assortment of planter boxes and pots; but when I stop by to see her, fatalism about the future pervades our conversation:

Temporary Oasis

Enveloped by leaves and bushes, Donata's house is a green oasis. For years I've walked past and seen her working there, crouched over seedlings with a couple of chickens rummaging around her. Usually we just say hello and go about our business, but sometimes we chat for a few minutes. When I complain that my beans don't grow as well as hers, she smiles momentarily and says I need to take better care of the soil.

Temporary Oasis (continued)

Then she lowers her voice and points to the construction site next door, where the bright orange arm of an excavating machine looks strangely out of place (figure 2.1). "They're making something there; I don't know what." Last year her neighbor sold his house to an investor, who has recently demolished it to make way for another apartment block. Keen to cash in, her children want her to do the same.

I try to offer empathy: "I hear them pulling down the old house every morning at 6 a.m.!" Donata raises gloved hands to her ears, grimaces, and shakes her head. "Yes," she agrees, "it's loud . . . many new people coming here now. Now I don't know people anymore." I refrain from criticizing the newcomers; after all, I'm one of them. Perhaps sensing my awkwardness, she resolves that "everything changes; it's like that, the world; it's just progress."

Research diary, Melbourne, December 14, 2023

Figure 2.1
Donata's garden, Moreland. Photo by the author.

One block away from Donata's house, Moreland's transformation is embodied in a strip of boutique restaurants on the iconic Lygon Street. References to the district's Italian heritage appear on the walls of these chic establishments in paintings of bygone Tuscan family farms, gray reprints of early migrants, obsolete coffee grinders, classic Vespa mopeds parked casually amongst the tables, and menus that promise traditional and fresh organic ingredients. As Moreland becomes increasingly inhabited by young professionals, such scenarios provide what Tim Butler (2006, 1) has observed in other gentrifying neighborhoods: "a ready-made narrative of icons from a past age suitably sanitized for busy lives."

These restaurants and bars have a similar gentrifying effect as the ostensibly sustainable apartment blocks that are reshaping Moreland's skyline. Both inflate land taxes, making the suburb unaffordable for older neighbors, who are forced to sell their properties to developers. Anguelovski and Connolly's international case studies of "the green urban life" and its associated real estate interventions are instructive:

> Interventions increasingly create new dynamics of exclusion, polarization, segregation, and invisibilization. Despite claims about the public good, these interventions take place to the detriment of the most socially and racially marginalized urban groups whose land and landscapes are appropriated through the creation of a "green gap" in property markets. (2018, 417)

The heritage narratives and gentrifying impacts of Melbourne property developers, estate agents, and restaurants contrast with those of emerging community projects committed to "learning from our productive past" (Gaynor 2018). Some of these have established common purpose with local governments and socially committed customers by drawing attention to the industrialization of food systems and the need for action to forge alternative futures. The two Melbourne cases that follow were among the projects surveyed when two Brazilian researchers visited to study the growing inclusion of urban agriculture in municipal policies. Funded by FAPESP, the visits included an initiative supported by Dr. Chris Williams to secure local government inputs, such as gardening equipment and salaries for horticultural staff, for food production in a public park (case one). The other Melbourne case study describes the CERES Fair Food initiative led by Chris Ennis, who has created an online platform to generate jobs for asylum seekers and market opportunities for small farms that protect productive green areas in and around the city.

The case studies show how producers, governments, and consumers can come together around a common cause. Williams and Ennis broker their cooperation by identifying mutually appealing goals associated with local food traditions. They then articulate these traditions as a basis for collaboration, demonstrating Putnam's (2007, 165) seldom-cited contention that inward bonding can be a "prelude" to outward bridging when effectively mediated. In the first case, from the east Melbourne suburb of Dandenong, Williams raises public awareness about food multiculturalism dating back to the establishment of the suburb's iconic farmers market in the mid-nineteenth century. In the second case, from the central Melbourne suburb of Moreland, Ennis uses community newsletters and education programs to recall lost connections with the natural environment and to encourage public adoption of informed food choices that push back against this loss.

Case Study: Food Traditions and Public Space in Dandenong

One-quarter of Australians were born overseas, and some councils, such as Dandenong in Melbourne's east, are especially diverse. The suburb's name derives from *Tanjenong*, or "lofty mountain" in Woiwurrung, the language of the area's First Nation Wurundjeri people. The arrival of British colonial settlers in the early nineteenth century displaced and decimated the Wurundjeri community, and by the 1850s, the entire area was converted to cattle grazing. Over the ensuing century Dandenong became an important hub for livestock trading, paper milling, quarrying, and brick manufacturing, culminating in a post–Second World War industrial boom that attracted migrants from Greece, Albania, and southern Europe. Now a rust-belt zone with high unemployment and a reputation for gang violence, Dandenong is home to some 156,000 residents from 150 different nationalities, more than half of whom were born in a non-English-speaking country (CGD 2017, 1,10).

Food options in Dandenong reflect the district's ethnic diversity, particularly around the central market that since the 1860s has been a hub of social life. Eager to leverage this diversity to promote healthy diets and combat high levels of obesity and diabetes, in 2013, Dandenong Council invited Williams to develop an urban agriculture project. The notion that cultural heritage might inspire residents to reconsider food choices was articulated in the *Greater Dandenong Regional Food Strategy 2015–2018*:

> Food is something that connects us all. It is a driver of employment and health and a significant part of our environmental, social, and cultural identity, . . . Greater Dandenong has the ability to have a very strong multi-cultural food identity providing great food variety that taps into the exciting flavours and food styles represented in the area. Greater Dandenong has a long history in all aspects of food. (CGD 2014, 2, 34)

In partnership with the horticulture team from the council's parks department, Williams proposed installing "edible landscapes" in public parks near Dandenong's busy central market to help achieve the Strategy's goals. His Novel Crops research project at Burnley Gardens had featured in national media (Backhouse 2016), providing a basis for selection of sweet potato varieties whose ornamental appearance would enhance visual appeal and whose yields would supply a local charity. These species and their cultivars are not widely available in Australian supermarkets and are rarely grown, yet they are a rich source of nutrients and common in many Dandenong residents' countries of origin.

In 2014, Williams and the council parks gardening staff created a sixty-square-meter edible planting bed in Dandenong Park. In-house horticultural staff have disappeared from most Australian councils since the 1980s, replaced by short-term contractors tasked with the basic maintenance of public landscapes. The Food Strategy supported the retention of horticultural staff through a promise to "stimulate stronger integration of food cultivation and production into urban and peri-urban locations" (CGD 2014, 38). The gardening team had already begun substituting ornamental flowers with lettuce, chilies, and corn for local charities six months prior to the sweet potato project. In its first harvest, in 2016, Dandenong park produced approximately 100 kilograms of sweet potatoes, taro, and yams, and in 2017, a plot in front of the market produced 110 kilograms.

As passers-by approached the garden beds in May 2017, Williams, his research student, our Brazilian colleagues, and I recorded seventeen informal conversations, some of which indicated familiarity with the crops and surprise at seeing them grow in Australia. More extensive interviews with eight residents revealed a perception that Dandenong's multicultural history favored flexibility in food choices and enthusiasm about the experiment.

Shola, who had migrated in 2015 from Lagos, Nigeria, wondered if the project might open opportunities for her community to get involved:

> I can't believe these [yams] are growing here. For us it's, like, a sacred food! Wait till I tell the Yoruba association about this. [laughs] They won't believe it. . . . We have a lot of festivals and parties and now I'm thinking, why don't we grow the yams ourselves because, you know, the markets don't always have them. (Interview, May 17, 2017)

Shola's enthusiasm was shared by other interviewees, and it contrasted with a concern raised in Dandenong Council that citizens may be opposed to the replacement of ornamental flowers with edible plants in public areas. The project's positive reception was reported in local media, encouraging the council to extend the range of edible crops and the land dedicated to their cultivation (Neill 2016). By 2018, community and school gardens had been revived or created in twenty-seven sites across the suburb.

When COVID-19 broke out, enthusiasts continued tending to the garden beds outside the market and in Dandenong Park. This attracted the popular television show *Gardening Australia* to feature it in a primetime episode about the benefits of food growing in small spaces during Melbourne's notoriously long lockdowns. By the time the pandemic subsided, volunteers from Dandenong's African community had been joined by friends from the nearby suburb of Pakenham, home of a grassroots initiative called the United African Farm. The growing momentum led Dandenong Council to re-engage Williams, who offered Burnley's soil and propagation facilities to assist the farmers. When I caught up with them during a potting session in November 2023, he explained his motivation:

> Imagine you're born in Nigeria and that you or your parents move to Dandenong. That would be hard, and you might have social barriers and identity issues. Now imagine that there's an opportunity to work with others in a wider community, plus university people and students. And even better, they're growing yams, which are connected with Nigerian identity in various ways. . . . I'm hoping this could be a first step for them to get training as landscape designers, or maybe people who look after trees, and other jobs. So I guess you could say I'm teaching garden literacy, so to speak. You asked me before if I think I'm an intermediary. I guess so, because what I do is put pieces of culture together in a kind of *ad hoc* way. It's a low-key experiment about using open space to show multiculturalism, and I'm hoping to develop this more in Dandenong by displaying foods from the places people come from. Shouldn't we ask how a postcolonial city like Melbourne can reflect people's historical memories and the plants and landscapes they know and miss? (Interview, November 2, 2023)

Figure 2.2
Harvesting in Dandenong Park. Photo by the author.

By putting "pieces of culture together in an *ad hoc* way," Williams demonstrates *bricolage* in action. One of these "pieces" was the preference of some residents for ornamental plants in public spaces, which he accommodated by selecting colorful and visually appealing edible crops. A second piece was the delight of other residents with yams and sweet potatoes as a critical ingredient in community festivals. A further piece was the local government's celebration of Dandenong Market as a historical site of fresh and diverse food that might inspire healthier diets. There was nothing necessarily compatible about these elements, but when combined in an "edible landscape," those who held them dear found common ground. By reinforcing bonds within each group, Williams provided a basis for outward bridges.

The project could not have advanced without Dandenong Council's progressive vision, which, in 2024, remains committed to funding staff and providing equipment to creatively remodel public land. The horticulture team has access to vehicles, tools, compost, and other resources that community groups often spend years raising money to purchase. The initiative, therefore, marks a departure from decades of state withdrawal from localized food production and resonates with earlier public employment and resourcing practices. Supportive policy frameworks constitute a key proviso for any intermediary, and the Dandenong project shows how such policies help to transform landscapes for collective benefit.

A second proviso, public background knowledge of urban farming, was less evident in Dandenong when the project began. The council's horticulture team leader, John Oakes, concluded that a "good argument" capable of galvanizing the multiple participants had begun to turn this around (interview, May 27, 2017). The argument had developed "organically," he said, through educational interventions such as information boards installed next to the planting beds, public feedback sessions conducted by two University of Melbourne research students, activities with local schools, and community "planting days." Williams secured council support for these activities, which generated knowledge about Dandenong's vibrant multicultural food heritage and the need for local government investment to sustain it. The resulting "argument," as John described it, was that the suburb's food multiculturalism should be conserved as a marker of local identity at a time of economic strain, proliferation of fast food, and related health impacts. Williams planned to deepen this argument by "displaying foods from the places people come from."

The Dandenong project demonstrates the importance of an enabling policy environment and an informed public, but it also reveals that the work of intermediaries is precarious. While culturally informed edible landscapes may provide a valuable community service, they are not equivalent to the research publications and grants demanded by most universities. During our discussion Williams reflected on this problem: "There was a time, a few years ago, when community outreach was all the rage. Back then universities funded engagement projects, but now the goal seems to be preventing any money being spent." Having survived a series of short-term academic contracts, Williams increasingly took on freelance consultancies with local governments and private customers. Professional stability is a key proviso for the sustained effectiveness of any food system intermediary, and yet a recurring finding of this book is that it often eludes them.

Case Study: CERES Fair Food

In 1982, the Melbourne council of Moreland sought expressions of interest to rehabilitate a 4.5-hectare waste landfill, which had been a bluestone quarry in the early 1900s and an Indigenous (Wurundjeri) hub for food gathering prior to European colonization (CERES 2023). An independent association of local organic food producers won the council-subsidized tender to establish the Centre for Education and Research in Environmental Strategies (CERES). Having survived (and in some ways thrived) through the COVID pandemic, CERES attracts 400,000 visitors each year and employs over 180 farmers, educators, and service staff. At the core of the organization's mission is public education about sustainable food production and practical methods for achieving it. Council grants and land allocation were critical for establishing CERES, but it has since become 95 percent self-funded through commercial operations. These include two markets that retail seedlings, plants, and organic produce; a restaurant; an events facility for conferences and workshops; a teaching service for local schools; an educational travel division; and most significantly the Fair Food organic home delivery service.

Fair Food's customers place orders online and collect their boxes from "food host" centers throughout Melbourne and its suburbs, attracting 985 members in 2012 (Chung 2012, 16). According to the service's director, Chris Ennis, by 2018, its gross income reached AUD $6 million, enabling it to employ fifty workers and serve over a thousand families and restaurants each week (interview, September 17, 2018). During the COVID lockdowns, home delivery orders tripled, generating around AUD $18 million and requiring Fair Food to grow its workforce to seventy-five (interview, December 6, 2023). These figures have since returned to their pre-COVID levels, and fresh food (including in supermarkets) remains more accessible in wealthier inner areas of Melbourne, whereas less advantaged peri-urban zones have a historically higher proportion of fast-food outlets and cardiovascular disease (Burns and Inglis 2007; Capone 2015).

Fair Food sources its produce from approximately 100 small farms in and around Melbourne. This has protected their land from takeovers by real estate developers and export-oriented wheat farmers, while facilitating long-term leases and tax concessions for productive use of green space. These outcomes align with Moreland's Food System Strategy (2017–2020) and its extension (2020–2024), which together commit AUD $395,000 to projects that promote fresh food growing, trading, and educating (MCC 2017, 2020). Moreland's mayor, Ms. Natalie Abboud, noted that CERES and its allies were "banging the drum" to protest the historical decline of horticultural green space, and that this influenced council's decision to adopt the Strategy (interview, October 30, 2017). As the first funded framework of its kind in Australia, the Strategy incentivizes the productive use of urban arable land and—as long as Moreland council maintains its progressive orientation—reduces the risk of small farms functioning as temporary placeholders for commercial or residential construction. The initiative's extension through 2024 takes the important step of employing a Food System Officer for four days per week to oversee its projects. Backed by this enabling policy proviso, Ennis relocated Fair Food to a larger warehouse in 2017, which by 2023 housed collaborations with a recycled clothing project and a sustainable timber initiative called Fair Wood.

Most of Fair Food's employees are refugees with uncertain migration status and few job opportunities, but every day they prepare lunch together, celebrate their cuisines, and share their stories. Like other asylum seekers around the world, their precarious residency and work status can be described as "liminal" (Tsoni 2016). Ennis believes that the risks this poses to Fair Food's stability is outweighed by their commitment and creativity:

> For many this job is their first anchor when they arrive in Australia, when they have no friends or family to connect with. We currently employ 26 nationalities, mainly from the Middle East and Africa, and you can see it on the walls in the murals they've painted and the music they play. A big part of building a stable work environment is simply making it fun. I've noticed that so many food warehouses are dry places devoid of culture, and as you can see, this place is different. . . . During COVID a lot of musicians lost their jobs, so we hired them to come and play here for the staff while they work, right there on that platform. It was an idea I had after visiting Cuba and learning about the people who read stories and the news to the staff who are making cigars. (Interview, December 6, 2023)

Victims of unjust circumstances and migration policies, Fair Food's staff are helping to build a more socially and environmentally just urban food system. Ennis (2012, 6) outlines the need for change in a CERES newsletter:

> In Australia in 1900 one in seven of us were farmers, today only one in 33 grows the food we eat . . . As we push our trolleys down the aisles of Coles and Woolworths, comforted by idealised rural farmyard pictures on our food labels, the reality is that our demand for cheap food all year round clears away forests and bush, erodes soils on millions of acres of once productive land, cages millions of

Figure 2.3
Chris Ennis at the Fair Food warehouse in 2023. Photo by the author.

> farm animals, poisons our rivers and oceans and overloads our atmosphere with carbon dioxide.

Noting how labels depicting the "idealised rural farmyard" divert attention from structural problems, Ennis uses the CERES newsletter, social media, and educational tours to raise public awareness about the historical trajectory of Melbourne's food system. Building background knowledge of endangered food traditions and the potential to restore them through social enterprise constitutes a key proviso for his connective work. Like Chris Williams in Dandenong, his efforts to protect land and livelihoods combine local government support with the power of private demand.

Online customers may not have space or time to grow food, but as public awareness about the historical trajectory of food systems grows, many wish to see the diversification of supply and demand for health, ethical, and other reasons. Fair Food and the wider CERES initiative do not seek to revive the past, but their educational services, employment model, and adoption of online technologies are building knowledge about the relevance of history to the future intersection of food and urbanism.

Feeding the Megacity: São Paulo

As in Melbourne, particular aspects of urban agriculture are on the rise in São Paulo, propelled by a unique blend of commercial, political, and historical pressures. Community farming cooperatives are proliferating, supported by the mayor's office (*prefeitura*), the University of São Paulo Urban Agriculture Study Group, and a wide range of national and foreign NGOs. These alliances are supporting local organic food production to ameliorate urban conditions while raising awareness about changes in land use, ecology, and political representation (Baptista da Costa et al. 2017; Giacchè and Silva 2014; Nagib 2016). Inspired by a 20 percent jump in national consumption of organics to USD $600 million between 2014 and 2015, 222 São Paulo municipalities began running urban agriculture projects in 2016 (Organics News Brazil 2017, Organicsnet 2016). The 2016 Connect the Dots initiative (*Ligue os Pontos*), whose impact among farmers is explored later, established an online platform for connecting producers and consumers, giving rise to the 2022 Sampa+rural information and subsidy system.

To service growing demand for fresh food and sustainable lifestyles, São Paulo real estate developers have begun building vegetable gardens and micro-orchards into their high-rise projects. They advertise green space as a premium feature that will revive citizens' long-lost connections with nature and with each other. Although this promise is as dubious in São Paulo as it is in Melbourne, we find evidence that some recent initiatives are more genuinely addressing the shortcomings that have historically eroded the city's ecology. To put this tension in context, it is useful to consider how São Paulo's urban conditions have changed over time.

When Prince Dom Pedro stood in São Paulo in 1822 to deliver Brazil's declaration of independence from Portugal, the city had become a commercial hub producing crops for local consumption, for trade with neighboring regions, and for national food security (Blaj 1998). At that time the city's southern, northern, and eastern zones were fertile floodplains well suited to agriculture. Small farms, or *chácaras*, in these areas provided vegetables and meat for the city's 30,000 residents and for Brazil's then capital Rio de Janeiro, while sugar production was concentrated further away from the city (Vieira de Melo 2006). As there was no legal distinction between urban and rural land, population growth gradually transformed farms into residential zones, pushing the *chácaras* outward to a food-producing "green belt" around the city's outskirts (Fernandes 2008).

The rise of coffee plantations in the late nineteenth century brought a new wave of European migration, nearly quadrupling São Paulo's population in one decade: from 64,934 in 1890 to 239,820 in 1900, and then to 580,000 by 1920 (Singer 1977, 47). Agriculture became organized into two concentric rings: the inner *chácara* belt characterized by small family farms and the outer *caipiras* dedicated to industrialized production of coffee, livestock, and sugar. The inner ring came under intense pressure as the automobile industry competed with rail transport in the 1940s and lobbied for highways across the city's suburbs and beyond. São Paulo's cultivated land decreased from 12,158 square kilometers to 9,536 between 1960 and 1970, as a dense network of businesses and factories developed along the new highways. Land beyond the municipal area was turned increasingly to export-oriented agribusiness, stimulated by a forced-draft federal industrialization policy (Pereira et al. 2012, 11).

Employment opportunities brought further migration, especially from rural areas, increasing São Paulo's population to 12.6 million by 1980. The rapid influx of people was met with a shortage of housing, intensifying the pressure for residential development on the city's fringes despite the lack of primary services. Disorganized and chaotic settlements, many of them illegal, encroached on forests, waterways, and the remaining *chácaras*. Economic stagnation in the 1980s saw a dispersion of São Paulo's industries to other cities, stimulated by federal incentives. The inner-city population began to decline, but the broader metropolitan population continued to grow, reaching twenty-one million by 2014.

By 2023, rural–urban migration had propelled greater São Paulo's population to nearly twenty-three million. The city's production of fresh food has dwindled to a few remaining family farms and waterways, which together constitute the last remnants of the green belt. A degree of legal protection has been afforded through the zone's inclusion in the larger (98,800 km^2) Atlantic Forest reserve (Morellato and Haddad 2000, 786). However, the area is under constant pressure owing to land speculation, illegal settlements, air and water pollution, illicit extraction of forest products, and highways that traverse ostensibly preserved areas. Fruits and vegetables cultivated in this area must travel 50–80 km to reach the city's main supply hub, the CEAGESP (Companhia de Entrepostos e Armazéns Gerais de São Paulo).

Servicing the city's wholesalers and markets, the CEAGESP trades an average of 283 thousand tons per month, mainly of fruit, vegetables, and derivative products originating from 1,500 municipalities in twenty-two Brazilian states and nineteen other countries. Industrial scale has enabled quantity, but mass production has compromised quality. In 2006 and 2007, researchers tested 240 products sold at the CEAGESP, finding that nineteen of these contained pesticides beyond the permitted levels and 28 percent contained unregistered ingredients (Gorenstein 2008). An aggregated study of nationwide tests between 2001 and 2010 found that 48.3 percent of food samples contained pesticide traces and 13.2 percent contained unauthorized ingredients (Oliveira Jardim and Caldas 2012, 607).

Brazil is the world's largest consumer of pesticides, using 914,000 tons in 2014–2015 at a cost of USD $9.6 billion (ABRASCO 2016). Despite growing evidence of negative health and environmental consequences, the federal government subsidizes and distributes pesticides—including glyphosate—to support the expansion of commodity crops (Nagib 2018). The approval of new pesticides has increased rapidly since 2016, when Dilma Rousseff was impeached and replaced by the conservative lawyer Michel Temer. Reliant on the agribusiness sector's political support, in 2019, the far-right Bolsonaro administration authorized 169 pesticides, twenty-four of which contain active ingredients banned in Europe (Phillips 2019).

São Paulo regulations such as Law 01-00891/2013 restrict the commercialization and use of nineteen pesticides, but compliance is low among farmers in the green belt zones, particularly those who do not claim to produce organic food. Among the consequences is contamination of the city's eight waterways, such as the Guarapiranga system that intersects the green belt and is responsible for 21 percent of the city's supply. One study identified thirty-one chemical compounds in São Paulo's water, many deriving from agricultural pesticides (López-Doval et al. 2017).

Urban sprawl has compounded water contamination, particularly in the Guarapiranga and Billings Basins, whose cumulative population has grown to two million people (SABESP 2017). In these and other supposedly green zones, overweight (including obesity) rates exceed the city's already high average of 54 percent (Ministry of Health of Brazil 2016, 5). These indicators reflect the lack of fruit and vegetable retail in peri-urban areas, underpinned by two centuries of suburban industrialization (Santandreu and Lovo 2007). The predicament demonstrates how fundamentally São Paulo's geographic and economic conditions have changed since the days of Prince Dom Pedro.

São Paulo's transformation cannot be reversed, but as in Melbourne, property developers invoke narratives about the capacity of traditional food production to restore lost connections between people and nature. "It's not only about the desire for healthy food," says one São Paulo consultant, "but also about a set of values that recognizes the importance of local production and its emotional connections" (quoted in Muniz 2017). This observation also holds true for other Brazilian cities, where one survey found that urban agriculture is stimulated by "respect and pride for a nostalgic cultural heritage" (Comassetto et al. 2013, 364).

Pursuing the market for this nostalgia, one developer of luxury condominiums in the upmarket São Paulo district of Itaím Bibi is installing vegetable gardens that enable residents to rediscover "the history of contact with the earth through urban agriculture" (Huma 2017). Another is designing exclusive apartment towers with "agro-forests" that "link aesthetics with functionality, restoring in a pleasant way the conviviality between people and the natural countryside" (Cardim 2017).

Restoring a quality of life lost to urbanization also characterizes the new central São Paulo Seed complex, whose eighty apartments each contain a three-square-meter patch of fruit trees. According to the developer, this feature adds ten percent to the value of each apartment because it represents "more than sustainability; it is a connection with nature" (quoted in Corrêa 2019). Another developer won the 2017 Master Real Estate Marketing Award for "hooking" customers on the promise to "bring the Atlantic Forest into the city" (Moreira 2017). Similar upmarket projects have appeared around São Paulo, offering vegetable patches and micro-orchards that extend the "benefits of nature" into urban zones (Ecoeficientes 2019), promote "conviviality" (Ideazarvos 2019), and enable customers to "pick fruit where it is least expected" (Grupo SP 2007).

As with the cases of "ecological gentrification" and "sustainability fix" analyzed by Gould and Lewis (2017), Dooling (2009), McClintock (2017), and While et al. (2004), the promises of São Paulo developers to restore lost human connections with nature and promote social conviviality do not address broader structural problems. These include the loss of productive urban and peri-urban land to real estate development, the contraction of funding for local governments to maintain green spaces, and lack of environmental protections against air and water pollution. Exclusive vegetable gardens atop luxury apartments reflect a larger green infrastructure trend that not only diverts attention from these issues but deepens what São Paulo urban planner Paulo Pellegrino calls "vegetation apartheid" (quoted in França 2011). Indeed, the city's "revitalization" demonstrates a form of gentrification fueled by the commodification of public space and "heritage conservation based on market demands" (Leite 2015, 175; Rubino 2005).

To this divisive backdrop, initiatives are nevertheless emerging in São Paulo to generate awareness about structural challenges and locally redress them. Gustavo Nagib and Angélica Nakamura (2020) trace these initiatives to the city government's creation of the Urban and Peri-Urban Agriculture Program in 2004, which gave rise to the Agroecological Cooperative of the Rural and Clean Water Producers of the Southern Region of São Paulo (COOPERAPAS). This in turn stimulated the formation of activist and research organizations including the Union of Community Gardens of São Paulo (UHCSP) and the USP Urban Agriculture Study Group in 2016. Although these groups pursue a range of agendas, Nagib and Nakamura argue that they each confront the common challenge of real estate expansion and promote urban agriculture to resist it:

> Real estate speculation is a great enemy of the quality of life in cities. Both in developed and developing countries, such speculation is to blame for the continuous horizontal expansion pattern in cities as a result of the absorption of farming land located in the outskirts by the urban body. In this scenario, urban agriculture is a resistance mechanism and one of the solutions to face urban crises. (2020, 3)

Community gardens are critical to this resistance, they explain, because they engender discussions and narratives about nutritional health and experiences of nature:

> Agricultural yield is secondary to the importance of discussing food production in urban spaces. . . . Based on organic and agroecological production, community gardens become collective trial labs for self-sufficient food production and reconnection with nature in the urban environment. The fight for a different model of agriculture and nutrition, one that respects natural cycles, becomes a motto for the community fight that helps define the agenda for local public policy demands in favor of agroecological agriculture. (2020, 4)

As in Melbourne, social bonds arising from the pursuit of food self-sufficiency and experiences of nature are at least as influential as economic goals. In both cities, articulations of food traditions that acknowledge historical injustices, including those that continue to afflict First Nation and refugee communities, constitute a powerful tool for consolidating advocacy networks. To this end, traditional food practices are invoked as a path to "reclaiming ancestral knowledge extinguished by industrial base monoculture" (Nagib and Nakamura 2020, 4). This is evident in two case studies from São Paulo that interrogate the nexus between government organic food programs and community aspirations.

The first of the two São Paulo case studies assesses the effectiveness of the Connect the Dots initiative, which was established in 2016 by the center-left former mayor to link under-resourced family farmers on the city's fringes with urban customers. In 2016 and 2017, I worked with the USP researchers who had previously visited Melbourne, Professor Thais Maud and Dr. Luis Amato, to interview seven producers at six farms and conduct informal conversations with customers and neighbors to explore their impressions of the initiative. Their experiences show that the online map underpinning Connect the Dots was a poor substitute for human mediation. Although in some cases the scheme provided equipment and storage facilities, its reliance on technological solutions overlooked the need for culturally sensitive interpersonal relationships. Follow-up visits in 2023, conducted with São Paulo food systems scholar Angélica Campos Nakamura, explored how intermediaries reemerged to rebuild social networks in the wake of COVID-19. While their human touch was widely valued, the pandemic eroded their already tenuous work conditions, forcing most to operate informally with little job security and constant risk of sanction by inspectors and police.

The second case study considers the long-term impacts of a project that had previously been carried out by my USP research partners to assess the impact of air pollution on urban crops. It summarizes that project's findings and how the researchers subsequently used these to produce a practical manual to help São Paulo urban farmers mitigate crop contamination. The case reveals the unintended consequences that can arise from a scientifically rigorous but socially detached evaluation of urban farming. It also demonstrates that inadvertent outcomes can provide valuable lessons, and even be turned to advantages, when intermediaries step in to build bridges between governments, producers, and their customers.

Case Study: Connect the Dots

São Paulo's expansion into surrounding green zones is a historical process whose impacts on environmental sustainability, social welfare, and dietary health have become acute. The 2014 City Master Plan, in effect until 2030, attempts to slow this process by reclassifying 25 percent of metropolitan land as "rural zones," prohibiting new residential construction and authorizing family farmers to access low-interest loans and subsidies for machinery and seeds (Prefeitura de São Paulo 2014, 18–19). The legal maneuver looks good on paper, but a year after implementation, 90 percent of São Paulo producers were still unable to secure technical or financial support (Ferreira 2015, 30).

Demand-side strategies have proven more successful, evident in municipal procurement of organics from family farms for public schools, hospitals, and enterprises. Underpinned by a price guarantee 30 percent higher than non-organics and a municipal ban on several pesticides, direct government purchases have stimulated a gradual shift to organic farming (Chatterjee 2016; Datamark 2016). Arising from the Master Plan, Laws 610 and 16140 aim to generate income for farmers while reducing obesity rates in peri-urban zones by requiring municipal schools to include agroecological produce in their meals (Camara dos Deputados 2015). In the southern district of Parelheiros, known for its peri-urban arable land but also for its low socioeconomic indicators and prevalence of obesity, farmers were supportive of the new policies. A total of 524 of São Paulo's 646 commercial fresh food farms are in this zone, whose distance from the city center (averaging 35 km) requires intermediaries to find urban customers and negotiate trade.

Interviewed in 2016, 2017, and 2023, six Parelheiros farm managers were eager to increase their urban sales but worried about the health effects of pesticides and the scarcity of organic options for their local communities. Like ten other producers we interviewed in the eastern districts of Cidade Tiradentes and São Mateus, they had tried to engage with government procurement programs that might enable them to expand agroecological production. Except for some limited success with school meals programs, their efforts had not yielded satisfactory results because—in the opinion of one farmer interviewed in 2023—"our government would rather spend money on the big export industries like beef and soybeans." Her contention puts fresh food farms in the context of expanding commodity plantations, whose comparatively favored territorial and political status seems inexorable as Brazil's trade with China booms.

Government attempts to support family farmers, such as a scheme developed by former Mayor Fernando Haddad (2013–2016) to expand the school meals program, face financial constraints. Attention has, therefore, shifted to the private sector as a potential source of demand that could sustain rural livelihoods while protecting peri-urban green space. In 2016, Haddad won the USD $5 million Bloomberg Mayor's Challenge for a project called Connect the Dots (*Ligue os Pontos*), which set out to create an online platform, a smartphone application, and logistical infrastructure to connect farmers on the city's southern outskirts not only with public schools and hospitals but also with inner-city private restaurants, supermarkets, and fresh food stores (Bloomberg Foundation 2016). Proposing to "find a solution to the urban sprawl threatening the region's local farmers and forcing them to struggle to make a living off their produce," Connect the Dots developed a narrative—amplified to the world by the Bloomberg Foundation (2017)—about combatting the negative consequences of the city's physical and industrial expansion. This agenda aligns with Article 5 of Law 610, which promises that "priority will be given to acquisition of organic produce directly from family farmers . . . traditional communities, and producers located in the proximity of public schools" (Camara dos Deputados 2015, 2).

The Parelheiros farmers we interviewed in 2016–2017 noted the challenge of building relationships with customers. To do this they relied on intermediaries (*atravesadores*, literally "crossers") like Roberto, who had grown up on a farm in the area and later had taken an entry-level job in the Ministry of Agriculture. Roberto blended his personal experience of fresh food production with professional knowledge of regulations and channels for promoting organic farming. "Officially speaking," he said, "my job is to gather data on production costs and yields, but it doesn't pay well. I spend more time making *arranjos* [informal arrangements] between farmers and the vendors in the markets [in central São Paulo]. . . . Nothing can be bought or sold if the products don't get to the markets." As the ministry has no official post for this work, Roberto operated informally and precariously: seeking suppliers and retailers, operating in cash, and avoiding tax inspectors.

Roberto's precarity was shared by Juliana, who worked full time as a transport and storage *atravesadora* for six Parelheiros producers. She had heard of the Connect the Dots scheme but, like Roberto, had not been invited to participate in it. Despite its focus on mediating the needs of farmers and consumers, the scheme had not generated employment or new opportunities for either. Juliana offered her opinion of the scheme: "Cutting out the middleman might save people money at first, but they'll see that a website can't negotiate good prices or deal with tax agents or find someone to fix a broken down truck . . . you need *jeitinho brasileiro* to get these things done." The Brazilian *jeitinho* is defined by Fernanda de Paiva (2022) for the Global Informality Project as follows:

> It is a personalised practice used to deal with extraordinary or unforeseen situations, and typically takes the form of a cordial approach (it is perceived to be more effective when accompanied by a smile or a gentle tone of voice). It involves a conscious act of bending or breaking the rules, and typically seeks short-term rather than long-term benefits. . . . [It is] a strategy to deal with the constraints of bureaucratic rules which make it hard for people to solve administrative problems that emerge in their day-to-day life.

Roberto and Juliana were proficient informal operators, bending rules to manage unforeseen situations in a way that Connect the Dots could not. Diminishing informality through a regulated digital platform is arguably one of the scheme's benefits, but amid the bewildering web of regulations that characterize São Paulo's bureaucracy, *jeitinho* permeates daily life and cannot be replaced.

During the *Lava Jato* corruption scandal that precipitated the impeachment of President Dilma Rousseff of the Workers' Party, in 2017, Mr. João Doria of the conservative Brazilian Social Democracy Party ousted Haddad and became São Paulo's new mayor on the campaign slogan "I'm a businessman, not a politician" (Magalhaes and Pearson 2017). Promoting business and connecting the dots between producers and consumers are potentially complementary propositions, but Doria's efforts to sell off USD $2 billion of public assets contrasted with the project's plan for subsidized storage warehouses and distribution hubs. A popular saying has it that Doria's agenda should have been called *desligue os pontos* (disconnect the dots).

After fifteen months in office, Doria resigned as mayor to contest the São Paulo state government elections of 2018, which he won. Vice mayor Bruno Covas assumed the vacated post and pursued a similar pro-business agenda, though less blatantly. He did not, for instance, advocate Doria's "modern management techniques," which among other things envisioned private advertising in public parks in return for donations of trees and seeds as a solution to the city's deficit of green canopy (Leahy 2017). However, Covas supported his predecessor's denationalization campaign, whose attempts to privatize public assets such as the Anhembi convention center and Interlagos racing track were broadly criticized (Seto 2017, 2018). Prior to his illness and death in 2021, Covas also favored new private real estate construction in and around the city, raising questions about the protected status of peri-urban land. His agenda has been carried forward by his deputy, another businessman-turned-politician, Ricardo Nunes.

Despite Haddad's replacement by three conservative mayors, Law 610 remains in effect and Connect the Dots has evolved into a broader online information platform and municipal subsidy program funded by the mayor's office. One Parelheiros farmer I visited in 2023, Ernesto, had secured funding from the program to purchase a small tractor and build a specialized tent for growing shitake mushrooms; another, Valeria, had won a grant to install a kitchen where school children could prepare meals with local produce. Both were receiving thirty hours per week of subsidized labor to assist on their farms through a branch of the municipal scheme called the Programa Operação Trabalho (Operations Work Program, or POT). A further legacy of

Connect the Dots is an online map of producers and their crops, now hosted on the municipal Sampa+rural website. Ernesto had been contacted by customers who had found him on the map, and he was pleased for their business, but standing beside a tall line of bamboo trees, he spoke to me of the resulting uncertainties:

> Sometimes I get calls from people who see me there [on the map] and want a certain vegetable or fruit, but often the information is out of date and I don't have that product. And since I don't know them, they have to trust that what I produce is good quality, and I have to trust that they'll come through on business side of it. So I don't really use the map. I find it much better to go to the Ibirapuera Fair [an inner-city farmers market] on Saturdays and meet my customers. They want to know about my background, so I tell them about my grandfather. He arrived here from Japan in 1907 and my family has been working this land with our traditional organic system ever since, . . . Look at these bamboos: their shoots provide food, but when they mature they provide strong fences, and we also use them as supports for the seedlings growing next to them. How we use space carefully is a big part of our tradition. (Interview, June 13, 2023)

Like Ernesto, Valeria found that a problem with Connect the Dots' online systems is that "there is nobody to negotiate with." Both seemed to lament the absence of personal relationships and on-the-spot solutions—that is, the *arranjos* and *jeitinho* described earlier by Roberto and Juliana. Social ties, they said, are important for finding customers, negotiating prices, and arranging transport, but they had also experienced problems with intermediaries who might furnish such ties. Ernesto found the prices charged by *atravesadores* too high and decided to invest in a truck to personally bring his produce to the Ibirapuera market. Valeria had also purchased a truck, but she ran into trouble when her friend, supposedly serving as an intermediary, began taking advantage:

> I asked a friend to make deliveries in the city and negotiate with my clients so that I could avoid contracting a driver, because drivers usually inflate their expenses. But it turned out he was using the truck for his own personal business with other producers, even on weekends. He denied it but I know because he got a speeding fine that came in the mail to me. Also, I heard that instead of finding new customers he was talking badly about the farm, so to avoid all this I'm going to sell the truck and just contract an *atravesador* who has his own one. (Interview June 13, 2023)

The risks that Ernesto, Valeria, and other southern São Paulo producers confront were intensified by the COVID pandemic. Despite some new uptake from private households and home-delivery services, the closure of schools, public offices, and private restaurants diminished urban demand for fresh food (Biazoti et al. 2021). Roadside truck stops also closed, with serious implications for a state whose extensive road network—noted earlier—reflects nationwide reliance on trucking to transport 61 percent of cargo (Samora and Mano 2020). To avert the economic consequences, in March 2020, the federal Ministry of Agriculture ordered highway diners and rest points across Brazil to remain open. Agribusiness exporters were protected by the order to ensure the flow of soybeans to China, but fresh food producers were not because the responsibilities of their intermediaries are a broader, from identifying customers to negotiating prices and physically presenting their produce at markets.

Roberto's former clients told me in 2023 that during the pandemic he had lost his job at the Ministry of Agriculture and no longer worked in the Parelheiros area. Juliana was still active, having diversified into the transport of farm equipment alongside fresh food. At one of the many "Graal" branded truck stops south of São Paulo, she explained that her work, like that of other intermediaries, was not legally authorized during the pandemic as an "essential service," and that consequently she had been unable to move freely. Her predicament substantiates the findings of a study conducted by the SMDU's Connect the Dots team during the first six months of COVID in 2020: Of 148 producers, the pandemic's impact was less severe for those who engaged in direct sales at the farm gate than it was for those who relied on inner-city markets. Moreover, its strongest impact was on the 57 percent of respondents who employed intermediaries to connect them with urban customers (Biazoti 2020, 5,7).

Connecting São Paulo's fresh food farms with public schools and private retailers requires more than truck stops and an online map of producers. Roberto's informal agreements (*arranjos*) and Juliana's workarounds (*jeitinho*) show that it is also a matter of building relationships. Yet the work of intermediaries in São Paulo and elsewhere remains unstable and precarious. COVID demonstrated the consequences of inadequate professional status as Roberto, Juliana, and other intermediaries found themselves unauthorized to travel and engage personally with clients. While the worst of the pandemic is over, growing reliance on digital platforms to mediate relationships is an ongoing problem. The chapter's final case study probes deeper into the dynamics of communication, illuminating efforts to mitigate the impact of traffic emissions on São Paulo's urban farms. The impersonal nature of scientific experiments, academic publishing, and mass media journalism demonstrates the need for human mediation between technology and tradition.

Case Study: Unearthing Food Safety

The impact of environmental pollution on urban agriculture is emerging as a public concern as São Paulo's urban farmers learn that their produce may be contaminated. The anecdote at the beginning of this chapter shows how coverage of this concern by a popular television news program distressed Genivaldo and potentially thousands of other urban farmers, who now face public and legal scrutiny (Globo 2016). The prospect of losing customers came as a blow to Genivaldo, but even more confronting was the reputational damage to a trusted farming tradition handed down by his ancestors.

The University of São Paulo experiment behind the news controlled for soil, water, and exposure time, concluding that airborne lead and cadmium particles associated with traffic density exceeded United Nations, European Union, and Australian safety limits in four of ten urban farms (Amato-Lourenco et al. 2016). The experiment was the first to show that heavy metals penetrate vegetables not only through their roots but also through leaf and stem pores and, therefore, cannot be simply rinsed off as typically assumed. It was a significant finding that deserves careful communication, but its hasty coverage on the evening news left little room for care or consultation.

The study's findings were foreshadowed more than a century earlier by Ebenezer Howard, who proposed in 1902 that electrically powered transport in garden cities would keep the "smoke fiend . . . well within bounds," allowing for cleaner delivery of food to markets (quoted in Parham 2016). As noted earlier, since the 1940s, the automobile industry has advocated for an alternative plan marked by the construction of new roads and roadside businesses over the city's green spaces. The latter argument found expression in former mayor João Doria's 2017 campaign to close bicycle paths installed under his predecessor, arguing that this would improve traffic flow and support small businesses. As the USP study suggests, the consequences of intensifying traffic for ecological sustainability and citizens' health are pervasive but largely unknown.

The morning after the news broke about the USP experiment, Chris Williams and I accompanied its authors to visit Genivaldo at his farm in the East São Paulo suburb of São Mateus. The meeting had been arranged to discuss the study's results with him and collectively consider measures to mitigate exposure to airborne contaminants. I recorded the ensuing encounter in my research diary:

Unmediated Media

The tone and timing of the media coverage has surprised the researchers, who now worry that although they had omitted the specific names and addresses of farms, the journalist's map of contaminated suburbs might leave Genivaldo and other project participants feeling betrayed. We pull up in the USP minivan on a dusty roadside shoulder, alongside a tin fence that serves as a wall for Genivaldo's property. Rapping on the rusting metal gate evokes no response, and fifteen minutes of phone calls and messages go unanswered. As we resign ourselves back to the van, the gate scrapes open and a woman clutching a mobile phone waves us in.

Behind the gate to the left, green banana trees reach upward alongside uneven gray cinderblocks, orange bricks, and the patchy grey cement of the neighbour's back wall. Stretching out in front, seemingly out of place in their ageing surrounds, lie vibrant garden beds—about 50 feet across and 500 feet long—planted with neat rows of spinach, lettuce, beans, and other leafy vegetables (figure 2.4). As with urban farms around the world, the paddock lies below high-voltage electricity transmission cables that keep real estate developers at bay.

Three men emerge from a door on the left, greet us with subdued nods, and invite us to sit with them on plastic chairs and tree stumps arranged in a rough circle. The usual salutations—handshakes, smiles, hugs, and kisses—are absent from this encounter. The USP team leader introduces Genivaldo, who begins speaking but stops as tears well in his eyes. The woman who had met us interjects: She says she is Genivaldo's wife, Sebastiana, and then explains that several customers have been in touch that morning about the news story. Genivaldo appears roused by her words, nodding in agreement as she speaks, and then he stands to deliver his message:

"You came here today but I haven't seen you for a long time. You know me and I've always been happy to work with you. But honestly, I didn't expect this. *Cara!* [damn!] I've been growing clean food—organic food—for forty years and my ancestors have been farmers for centuries and last night the news said my crops are unsafe. Now my customers don't trust me! This place is special to me and we have a reputation. So now what will happen?"

Research diary, São Paulo, June 17, 2016

Figure 2.4
Genivaldo and Sebastiana's farm in 2023. Photo by the author.

The conversation that ensued conveyed the researchers' regret about the unintended consequences of the media coverage and their commitment to working with Genivaldo on practical steps to mitigate contamination. Some of these steps were foreshadowed in the original publication, which had shown empirically that distance from traffic and protection behind natural barriers reduce chemical uptake, and that trees are especially effective for blocking and absorbing harmful particles.

Other remedies, such as replacing leafy vegetables with root crops in areas near streets, were formulated with Genivaldo and other study participants over the subsequent months, leading the team to produce a manual of best practices for circulation among urban farmers, planners, and policymakers. Among its recommendations were that crops be positioned away from dense traffic, that thick bushes encircle leafy vegetables, and that farmers prioritize root crops when neither of these options is possible (Amato-Lourenco 2018). Sebastiana invited the local cooperative, the Association of Eastern Zone Producers (AAZL), to participate in the manual's production and ensure that the farm's interests were protected. Through AAZL's network, she corresponded with nearby farmers and local officials, gradually becoming a point of contact between these groups and the USP researchers. Each was asked to share the manual's insights in brochures, information sheets, and social media.

When I returned to the farm in 2023, five years after the manual's publication, Genivaldo had passed away and Sebastiana had become its manager. With the help of two employees subsidized by the municipal POT scheme, the farm was alive with green, yellow, and pink nutritional and medicinal plants. Sebastiana offered me a mug of garden-grown *melissa* tea and updated me on the intrigues of the intervening years, including her neighbor's attempt—outlined at the chapter's outset—to acquire her land:

> After the news said our farm was contaminated, we had so many calls and questions from worried customers. *Meu filho*! With our neighbor trying to take our farm from us, all those *fofocas* [gossip] about chemicals in our crops traveled fast and we nearly lost us our lease. But look at these beautiful banana trees along the wall. Bananas have always been special to me because my father grew them when I was growing up in Pernambuco, where Genivaldo and I met. I keep my memories of both of them in all these banana trees, which are also protecting us from contamination. People know me and this farm for my family's traditions, and I hope they will for many more years. (Interview, June 12, 2023)

Sebastiana's reputation as a long-term champion of organic farming proved valuable to the defense of her *comadato* lease. With AAZL's support, character references and statements were gathered from the farmers, officials, and journalists who had previously been involved in the manual's production and dissemination. AAZL's coordinator, Regiane Nigro, offered her perception of recent events:

> His [the neighbor's] timing was planned. Genivaldo was gone, and the Globo [news] report got the public worried about the farm's quality. We've talked a lot about it in the cooperative. . . . We concluded that the problem was timing. A scientific experiment has relevance at the time the experiment is done, so if there's a problem with air quality from a construction project or something then it gets detected. But journalists report these things like they're permanent problems even if they're temporary. These interactions were all short term so didn't communicate the reality, but now we're more careful that people are involved in our members' projects who can tell the whole story. (Interview, June 12, 2023)

While the news report had identified and circulated a scientific finding that was relevant to the public, it lacked the nuance and responsiveness of personal communication. Like the online map that emerged from Connect

Figure 2.5
Dona Sebastiana. Photo by the author.

the Dots, the mass media's detached transmission of sensitive information stripped away the local sensibilities and personal motivations that characterize the work of human communicators.

Observing the transmission of tradition in the previous cases illuminates how encompassing conditions shape food systems but are simultaneously shaped by bottom-up activism. Where governance structures, public knowledge, and employment opportunities were weak, intermediaries actively strengthened them to advance their cause. The chapter's final section suggests that this dialectical process of adapting to and shaping the three provisos is central to the work of food system intermediaries.

Conditioning the Preconditions

As urban agriculture gains prominence around the world, real estate developers and community activists are articulating food traditions in distinct ways. This can be seen in Australia and Brazil, nations that share a comparable history of colonial land use, dependence on commodity exports, and industrialization of food production and consumption. Growing interest in urban agriculture manifests in the real estate sector through exclusive condominiums featuring vegetable patches and fruit trees, advertised as a means to restore citizens' lost connections with nature and each other. Sold at premium prices, these features ostensibly "draw locals together" and enable them to revive "the history of contact with the earth" (Cardim 2017; Gardiner 2017). The "really old-style communities" that supposedly promote "sustainable eating" in Melbourne share the exclusive appeal (and elevated prices) of São Paulo's mid-city "agro-forests" (Cardim 2017; Farrelly 2018; Ideazarvos 2019; Lindsay 2016). Such language invokes ecological heritage but obscures the loss of enabling conditions such as protection of productive land, subsidized horticultural workforces, and environmental stewardship.

Land use is especially contested, with urban agriculture potentially contributing to "vegetation apartheid" as green zones become unaffordable and exclusive (França 2011). Cultural and economic diversification, pledged around the world in urban policies to achieve a better "social mix," tends to be temporary in such gentrifying contexts (Bridge et al. 2014, 1133). The narratives employed by developers are, therefore, incomplete, recalling the benefits of diverse food systems but forgetting the need for supportive structural conditions. In the case studies, intermediaries articulated more complete narratives about urban agriculture. Their projects identified the need to push back against systemic deficiencies, leveraging progressive municipal urban food policies—a key enabling proviso for intermediaries—while exposing the shortcomings of detached technological solutions.

In the culturally diverse Melbourne suburb of Dandenong, the municipal government aspired to diminish high levels of obesity and diabetes by promoting healthier diets. The "novel crops" work of Chris Williams provided an interface between residents' food growing traditions and the government's health-oriented agenda. School interventions, community feedback sessions, and information boards raised awareness about the district's vibrant multicultural food heritage and the need for workforce planning and creative land management to conserve it. Permeating these activities and cutting across ethnic identities was a concept of food that, as the Council's strategy document put it, "connects us all."

Connecting distinct communities was also a priority for CERES Fair Food, which harnessed growing urban demand for organics to help family farmers earn an income and retain productive land. Using social media, newsletters, and educational tours, its director, Chris Ennis, urged reflection on the social and environmental consequences of "cheap food all year round" (Ennis 2012, 6). Like the Dandenong project, Fair Food educated its customers about Melbourne's First Nation origins, the contributions of asylum seekers, and the need to support food producers as proliferating real estate and expanding commodity plantations transform Australia's fastest growing city. Ennis articulated these themes to personally engage school groups, customers, and farmers. This ensured a human touch behind what might have otherwise become an impersonal online platform.

São Paulo's Connect the Dots initiative set out to confront decades of urban sprawl by linking peri-urban farmers with inner-city markets. To advance this goal, the initiative was awarded a grant from the Bloomberg Foundation to develop an online map and information platform. Digital mediation can offer efficiency gains while potentially diminishing the corruption of *atravesadores*, but shortcomings emerge in the absence of face-to-face communication. As Roberto and Juliana demonstrated, informal arrangements (*arranjos*) and on-the-spot workarounds (*jeitinho*) require relationships forged by real people with local knowledge. The COVID pandemic diminished urban demand for fresh food as schools and restaurants closed, and exposed the precarious work conditions endured by São Paulo's intermediaries.

Finally, USP's experimental testing showed that airborne lead and cadmium now join sugar, pesticides, and other industrial inputs to undermine the quality of urban diets. The discovery of these contaminants in farms around São Paulo brought to light the city's historical reliance on cars, resulting air pollution, and the need for consultation with food producers and policymakers about remediation. The study's portrayal on the evening news was less reflexive, alerting the public to safety concerns but damaging the reputation of urban farmers like Genivaldo. Recognizing the unintended consequences of their work, the USP researchers followed up with Genivaldo and Sebastiana to formulate techniques for protecting crops and to create a manual of best practices for farmers, policymakers, and researchers. After her husband passed away, Sebastiana drew this nascent network into culturally informed conversations that were beyond the scope of television news, websites, and other mass media.

While intermediaries in the cases presented local knowledge as a source of insight for redressing the erosion of urban food systems, their ability to do so was influenced by political conditions. Dandenong Council's in-house horticulture staff, Moreland's Food System Strategy, and São Paulo's subsidies provided the administrative architecture—a key enabling proviso—for their activities. In the context of state-level and national-level legal support for agribusiness exporters and property developers, the longevity of such progressive local policies is uncertain. To defend their cities' urban farms, CERES in Melbourne and Connect the Dots in São Paulo relied not only on state inputs but also on emerging private markets. The disposition of intermediaries to simultaneously engage with the public and private sectors enhanced their ability to establish common interest in improving historically corrosive structural conditions.

Cultural provisos were at least as influential as political and economic conditions. As noted in the book's introduction, the traditions invoked and refashioned by food system intermediaries resonate stronger when the public is already familiar with them because, in Marshall Sahlins's words, "an invention of tradition entails some tradition" (1999, 417). In the east Melbourne suburb of Dandenong, Williams used school incursions and information boards to recall the suburb's multicultural food history. Meanwhile, in Moreland, Ennis highlighted Melbourne's farming heritage and loss of connection with the productive process in newsletters and education programs. In eastern São Paulo, Sebastiana's reputation as a champion of traditional organic farming was central to the legal defense of her land from a neighbor determined to acquire real estate. And in the southern Parelheiros district, Roberto's *arranjos*, Juliana's *jeitinho*, and Ernesto's explanations to customers about his farm's organic history demonstrated the benefits of culturally embedded knowledge over a digital platform.

The contributions of intermediaries and their networks in this chapter's scenarios did not translate into professional stability. The sweet potato project in Dandenong, mediated by Williams, yielded social and environmental benefits but did not produce the research publications and grants that universities expect of academic staff. Challenged to define the project's merits for increasingly meticulous annual performance reviews, he had to reorient his work toward freelance contracting. In Moreland, Ennis built Fair Food to mediate between peri-urban farmers and inner-city customers. While the project enacted its social values by employing asylum seekers, their insecure rights to work and residency were a constant source of personal anxiety and institutional uncertainty.

The COVID pandemic exposed the professional precarity of São Paulo's intermediaries when lockdowns impeded travel and face-to-face meetings.

An executive order designed to protect the export soy industry kept Brazilian truck stops open, but the needs of fresh food intermediaries to travel, negotiate, and personally present crops to customers were overlooked. Insecurity was also a daily stress for Dona Sebastiana, whose freehold territorial lease (*comadato*) left her, like other São Paulo producers, vulnerable to the priorities of corporate landholders and competitive neighbors. Under these conditions, farmers' cooperatives provide an important support framework when journalists—and academic researchers—come knocking.

When adequately supported, intermediaries offer valuable answers to societal questions about food's relationship with land, livelihood, and ecology. The easy reconciliation with nature offered by planter boxes atop expensive apartments cannot answer these questions. Solutions are more likely to arise from projects that draw attention to long-term structural shortcomings, offer localized adaptation strategies for urban farmers, and develop policy recommendations that link present challenges with historical circumstances. Nagib and Nakamura argue that such cross-cutting projects require community networks forged from the bottom up (2020, 6):

> Activists and farmers have established new, closer relationships, as well as social and spatial bonds that aim at changing intra- and peri-urban area spaces, as well as the mechanisms of institutional politics. . . . This results in the establishment of new networks and manifestations that transform the city and social relations without the need to wait for bureaucratic and institutional politics.

Encouraging the formation of "social and spatial bonds" is a core function of food system intermediaries. This chapter has shown that they advance such convergences by providing city officials and consumers with historically informed narratives that identify the need to invest in productive green space, employment, and environmental conditions. Articulating these narratives establishes a basis for consensus and cooperation, and for contesting less reflexive commercial appropriations of the past. Underpinning this process in the cases was a portrayal of urban agriculture not as a purchasable personal commodity but rather as a living collective tradition. These diverging interpretations resemble a broader global conflict over the meaning of food, described by Molly Anderson (2024, 45):

> The meta-narrative of food as commodity is promoted by people and organizations with a financial stake in current food system operations, such as agribusiness and the WEF, and thus somewhat suspect. On the one hand, they have a strong interest in accurately assessing risks and appealing to end-users; but on the other hand,

> they have an incentive to keep in place ways of doing business from which they are profiting, even if these cause damage to the public good.

The chapter has found that visions of food's economic and social functions diverge as widely in urban farms as they do at the scale of commodity plantations. Like the scenarios visited in this chapter, those that follow show intermediaries working as agents to consolidate local farming projects as their cities expand and grapple with access to healthy food. Balancing personal visions with collective values becomes more difficult as their networks grow. This is apparent as two intermediaries, discussed in the next chapter, attempt to create jobs in marginal zones of Beijing and Rio de Janeiro. In addition to defending community farms from developers, politicians, and (in the case of Rio) a drug cartel, they must build trust between low-income rural migrants and high-income urban consumers. Under these circumstances, converting inward bonds into outward bridges becomes a practical need that challenges their leadership and livelihoods.

3 Sowing Seeds of Trust in Beijing and Rio de Janeiro

I couldn't tell if Julio was angry or excited as he boomed from behind the principal's closed door. This was the last school visit in an exhausting series of meetings that 2016 summer day in our quest across Rio de Janeiro to build support for the municipal government's *Hortas Cariocas* urban farming initiative. When he invited me the day before to visit some of his projects, I had imagined a casual trek through a few of the program's fruit and vegetable gardens. Reality dawned at 8:15 a.m., when we left the Municipal Secretary of Environment to assemble a delegation of school policy advisors at the Ministry of Education. There was nothing casual about the ensuing negotiations, explanations, and attempts to find common ground among environmental protection officers, education officials, and school administrators.

Where would Julio's proposed garden fit within the school grounds, who would arrange the required teacher training in horticulture, and what incentives would the school be offered for participating in the *Hortas Cariocas* program? Such contentions generated fiery discussions at each of the seven schools and communities we visited across the city that day. A journalist researching the program told me that this was a normal day, except for my presence. I had been invited, she revealed, because bringing along a foreign professor would strengthen Julio's argument that organic fruit and vegetable production is critical for generating incomes, reviving lost connections to land, and educating publics about food traditions in and around cities worldwide.

Julio aligned his agenda with the advice of the United Nations Food and Agriculture Organization that governments should fund urban and peri-urban agriculture as "a key component of robust and resilient urban food systems" (FAO 2014, i,84). In his words:

> Community gardens won't solve global hunger, but they can provide jobs and a real response to a global problem. Some people support them for environmental reasons while others do it because they want organic diets. Some have political motivations while others do it simply to provide food for themselves and earn a living. But when you build relationships and understanding between these groups, the sum of the parts is bigger than the pieces. (Interview, November 21, 2015)

These arguments are salient in Brazil and other countries where industrial soybean plantations, expanding to meet growing demand from China, are encroaching on fresh food farms and forcing an exodus of rural residents into cities. In 2022, soybeans were again Brazil's top export (worth USD $47 billion), fueled by the quadrupling of soy plantations to forty-four million hectares between 1990 and 2022 (Ozbun 2023; Schnepf et al. 2001, 41; UN-COMTRADE 2024; Zalles et al. 2019). The US Department of Agriculture projects that to service Chinese demand, Brazil will convert a further twenty million hectares to intensive cropping by 2030, representing "one of the fastest rates of cropland expansion in the world" (Valdes 2022). The World Wildlife Fund's (2021) warning that "the soybean industry is causing widespread deforestation and displacement of small farmers and indigenous peoples" reflects Brazil's status as the world's leading exporter of the protein-rich grain and biggest importer of chemical pesticides and fertilizers (OEC 2021). Citing statistics like these to argue for a more diverse and balanced food system, by 2023, Julio had secured municipal funding for urban farms and associated salaries in thirty-two communities and twenty-five schools, producing eighty tons of fruits and vegetables each year. As in São Paulo (chapter 2), the importance of such community projects became evident during the COVID pandemic, when food shortages in Rio's most disadvantaged suburbs intensified, generating mainstream press coverage of the largest Hortas Cariocas garden in the district of Manguinhos (Barbon and Cheibub 2021).

Two hours on, Julio's debate with the school administrators showed no sign of winding up, so I escaped "to the bathroom" for a moment of peace. Exposed light bulbs outside the principal's office, irresistible to circling moths and other nightlife, dimly lit a noticeboard featuring students' designs of sustainable cities. Meticulous plans for apartments lined with vegetable gardens and schools decked with solar panels offered answers to the question glued above in green cut-out letters: "*Quais são as grandes questões para o nosso futuro?* (What are the big questions for our future?)" Casting my mind back across the day's visits to inquisitive classrooms and cautious staffrooms,

I recalled the question posed by a teenage student: "Why does Brazil need to produce so many soybeans?" The beginning of Julio's answer, "It's a little complicated. . . . ," provoked memories of life with the Wang family and its love of pork in Beijing (chapter 1).

Nutritional safety and security have become pressing concerns in Beijing and Rio, and initiatives are gaining pace in both cities to build more sustainable and localized food systems. Emerging alliances between local governments, community associations, and nascent private businesses are replete with tensions, and their viability is conditioned by the disposition of distinct interest groups to work together. As chapter 2 showed, food system intermediaries try to build trust between such actors by blending traditional farming practices with political diplomacy and business acumen. This chapter observes their impact among the people they employ, most of whom have migrated from rural areas that no longer accommodate their expertise. On Beijing's outskirts, their small farms have been converted into new residential suburbs, while in Rio's neighboring states, their towns have been engulfed by agribusiness plantations dedicated increasingly to soybeans. Unlike millions of others forced to make new lives in the world's growing cities, these farmers count themselves lucky to find urban jobs that harness their knowledge of traditional agroecological practices.

The chapter's first section considers the need for better synergies between government, business, and civil society in China and Brazil, where for centuries emperors and politicians have staked their credentials on the reliable supply of food. Deepening agribusiness relations between the two nations reflect the industrialization of agriculture, which is provoking similar concerns in both about nutritional health, ecological impact, and social welfare. Bottom-up alternatives are, therefore, emerging to restore trust in their food systems, including community farms in Beijing's urbanizing northeastern zone and in central Rio's Manguinhos district.

The second section offers case studies of these projects, focusing on the attempts of intermediaries to reconcile government policies with community priorities and rural food traditions. Their work confirms the importance of enabling provisos, such as policy support and public familiarity with urban farming, while further illuminating the intersection of bonding and bridging. The cases suggest that the consolidation of inward solidarity to provide a platform for outward collaboration is conditioned by the dynamics of societal trust. Overcoming public concerns about food safety in Beijing and healing fractured state-society relations in Rio demonstrate that trust—and its absence—is critical to the positive-sum integration of internal and external interests.

The final section finds that the efforts of intermediaries to build collaborations make them valuable to their communities but also vulnerable to collective overdependence on the connections they forge. This vulnerability is deepened by insecure and unstable professional conditions, a problem also endured by intermediaries in chapter 2. In this chapter, they reflect on these conditions in Beijing and Rio, identifying the need for contingency plans should their work be impeded by politics, health, or jealousy.

The Politics of Trust

"Societies," writes sociologist Nan Lin, "must have consensual rules and collective trust for them to function" (2001, 148). Applying this observation to China and Brazil shows that consensual rules and collective trust have evolved differently around the world.

Chinese commentators often argue, with varying cogency, that their nation's history reflects a Confucian tradition that prioritizes collective advancement over individual benefit. This orientation, they say, is evident in the pursuit of consensual "harmonious" development, "holistic" outcomes, and trust in state stewardship over national and international affairs (Pan 2004; Wen 2004). Born in 551 BCE, Confucius is no stranger to appropriation: He was celebrated as a model of filial piety and imperial loyalty under successive ruling dynasties from the Zhou (770–256 BCE) to the Qing (1636–1912), rejected as an antiscientific superstition during Mao Zedong's Cultural Revolution (1966–1976), and revived as a symbol of national pride after 2000. His recent reappearance is sustained in part by a justification of state authority as an ancient tradition embedded in imperial heritage, but also by visions of Confucian democracy that constitute "critical and constructive ideas relevant not only to China, but to the wider world" (Bai 2023, 230).

China's growing global impact has reawakened a younger but equally powerful political tradition in Brazil. Colonization by Portugal and subsequent domination by US business interests generated deep sensitivity about dependency, which, by the 1970s, developed into a suite of policies, particularly in the agriculture sector, to defend national sovereignty (Cardoso and Faletto 1979). Invoking this defiance, Brazilian critics now complain that China's hunger for their nation's commodities constitutes a form of neocolonial dependency (see chapter 5). The increasingly mechanized production of raw soybeans, they say, neither adds value nor generates employment.

Sugar, coffee, and livestock estates have long characterized Matto Grosso, Goiás, Minas Gerais, and other Brazilian states, but the recent expansion of soybean plantations is different. Encroaching on small farms that once provided cities with fresh food, they have provoked an exodus of young people from rural towns and accelerated urban sprawl, putting further pressure on peri-urban green zones. As Brazil's cities grow, unequal access to basic jobs, food, and green space between richer and poorer suburbs has deepened the fragility of societal trust and, as seen chapter 2, provoked what urban planner Paulo Pellegrino calls "vegetation apartheid" (quoted in França 2011). A 2023 study by the Getúlio Vargas Foundation concludes that the richest 10 percent of Brazilians earn 42 percent of the nation's income, while the poorest 20 percent earn only 3.3 percent. When factoring in data from tax returns, this generates a gaping national GINI coefficient of .71 that divides 217 million people economically and racially (FGV 2023; CEIC 2023).

Professor Wen Tiejun of Renmin University has argued that Brazilian urbanization provides insights for China, which he says must avoid the negative aspects of "Latinamericanization." For instance, Brazil's sprawling *favelas* have engendered marginalization, poverty, and delinquency, which in China would slow economic growth and threaten political stability (Wen 2005). While Chinese policies have averted such extreme urban inequality and its consequences, nutritional health indicators in both contexts reveal disturbing similarities. Brazil exhibits a clear correlation of low income with rising obesity, diabetes, and related cardiovascular complications (Governo do Brasil 2017; European Society of Cardiology 2016). A similar phenomenon has emerged in China, where heavily processed diets, fast food, and limited access to fresh produce have generated health risks that are attracting the attention of medical advisors and community associations (Cheng 2011; Liu 2017; World Health Organization 2016). Concerns about food safety and security are becoming more prevalent in Chinese and Brazilian cities, undermining public trust in their food systems.

Case studies from Beijing and Rio find that innovative projects are emerging to improve nutritional health in underserved communities while servicing broader public demand for organic food. These scenarios suggest that productive forms of state support for localized food cultivation are enabled by—and, in turn, enable—trusting relationships between producers, consumers, and municipal governments. Critical to these junctures of cooperation are intermediaries who are familiar with the values and priorities of the communities in which they operate and who, therefore, possess a sound understanding of the conditions for trust in each. Shi Yan in Beijing and Julio Barros and Fabiola Corrêa in Rio are intermediaries who fit this description, linking the economic agendas of municipal authorities with local experiences of land stewardship and global concerns about hyperindustrialized agriculture.

Beijing and Rio face food system challenges that reflect different scales, degrees of socioeconomic inequality, and structures of governance. As Chinese urbanization gains pace, Beijing has witnessed rising public concern about the safety of produce on supermarket shelves, stimulating customers to seek out and establish relationships with organic farmers on the city's periphery. Rio has seen peri-urban farms disappear as industrial agriculture escalates migration from provincial towns and exacerbates urban sprawl, provoking neighborhood actors to introduce inner-city agriculture programs in partnership with local governments. In both contexts, the consequences of hyperindustrialized agriculture are driving attempts to build alternative food systems through new combinations of public, private, and community inputs.

The deepening need for local alternatives to industrial food raises the question of appropriate scope and scale of government, corporate, and civil society inputs. The experiments underway in Beijing and Rio show that states, markets, and communities all have a role to play, and that the encompassing need for safe and healthy food generates opportunities for dialogue between these sectors. If government intervention is to strengthen trust across sectoral lines rather than "substitute" it as Fukuyama (1995, 27) cautions, then it must create enabling policy conditions for collective action. Intermediaries help to activate these conditions by teasing out the complementarity of each group's distinct capacities, or "resource structures" in Nan Lin's (2001, 39–40, 67) terms, and harnessing these behind a common goal. Applied to food systems, Lin's "resource structures" resemble the provisos for effective intermediaries discussed in previous chapters: supportive policy settings (e.g., land titles and subsidies for emerging enterprises), public familiarity with unifying food traditions, and stable working environments. Healthier and more sustainable food systems require relationships that bring these administrative, associational, and professional qualities into what Theda Skocpol calls "close symbiosis" (1996, 20).

As discussed in the introduction, symbiosis across sectors entails mutually supportive bonding and bridging, hinted at by Robert Putnam (2007, 143–44, 165) and evident in chapter 2. Food system reformers who try to balance internal community loyalties with external cooperation face a provocative question: "Do the high levels of integration characterizing indigenous social relations in many poor communities actually constitute a resource that can be used as a basis for constructing substantive development programs? If so, how?" (Woolcock 1998, 181). The projects discussed in this chapter demonstrate the complexities of integrating community social resources with official policy priorities and the role of food system intermediaries in building trust not only between specific actors but more broadly across society. Together the cases show them reaching beyond their communities into national and international affairs, extending themselves in valuable yet precarious ways.

Feeding the Masses in Beijing

For 500 years prior to the 1911 overthrow of the Qing dynasty, Beijing's Temple of Heaven hosted a ceremony of paramount importance for the consolidation of state-society trust, and food was at its core. Twice each year, the Son of Heaven, as the emperor was known, petitioned for the empire's prosperity in the Hall of Prayer for Good Harvests (祈年殿). As an act of mediation between heaven and earth, the ceremony's public message was clear: Entrust your food security to the supreme authority of the emperor. Divine intercession was backed up with earthly intervention, including sustained investments in flood protection and grain storage (Li 2007). Mencius, a disciple of Confucius, had argued that subjects were entitled to rebel against the emperor if his heavenly mandate were compromised. Productive harvests were taken as evidence of the emperor's divine credentials and were, therefore, critical for averting social disorder and maintaining public trust in the political system (Cullen 2017, 30).

After seizing power in 1949 and establishing the People's Republic of China with Beijing as its capital, Mao Zedong faced the perennial challenge of ensuring food for a growing population. Public confidence in his Communist government's ability to do so was shaken by the Great Famine

(1958–1961), whose contrasting mortality estimates (between three million and fifty-five million) demonstrate the tragedy's political sensitivity to this day. The famine's causes include over-reporting of grain stocks by fearful provincial officials, outdated production and distribution methods, and a policy that forced millions of peasants to abandon their farms for iron and steel factories (Wemheuer 2014). Officially the episode has since been named the "The Three Years of Natural Disasters" and reinterpreted under President Xi's "Mass Line Campaign" to assert the need for comprehensive government control over food security.

Having grown from thirteen million people in 2006 to twenty-one million in 2023, Beijing's urban population continues to pose challenges for the sufficient supply of clean and sustainable food (CPDRC 2008; PopulationStat 2023a). The farm villages that surround the city, which have long supplied the urban center, have experienced an exodus of young people and diminishing capacity to safely produce fresh food. The so-called left-behind villagers—typically children and people older than sixty years—are burdened with growing workloads and are neither equipped nor trained to enforce quality control procedures (Ye et al. 2012). Although large producers, distributors, and supermarkets have been responsible for a series of notorious food safety breaches, family farmers are more vigorously targeted by inspectors. This "scale bias" has intensified with economic liberalization, leading Zhang and Qi (2018, 116) to conclude that "market-based reforms in China have disembedded food production from social relations of trust and reciprocity."

Consumer concerns about food safety have been stoked by scandals including milk powder and animal feed found to contain the plastic compound melamine, pork laced with the muscle growth steroid clenbuterol, and spoilt beef imported from Brazil laden with preservatives (Barboza 2008; Condon 2017; Qiao et al. 2012; Smart and Smart 2016, 99). These and numerous other cases of contamination have stimulated public consumption of supposedly healthier processed and canned products, often imported from Hong Kong, Australia, and elsewhere.

Greater reliance on processed food has brought its own problems. A study by the Chinese Center for Disease Control and Prevention found that 33 percent of Beijing citizens are overweight (with a body mass index between 25–30) and 26 percent are obese (body mass index above 30; Liu 2017). Across China, overconsumption of sugar and fat, compounded by physical inactivity, is fueling a previously unseen rise in diabetes to nearly

10 percent of the population (World Health Organization 2016). Type II diabetes, which comprises 90 percent of these cases, is directly linked to dietary health.

China's urban consumers are becoming more concerned about dietary health and food safety, instigating a quest for trustworthy local sources. "Middle classes in big cities," write Zhang and Qi (2018, 118), "are beginning to distrust official food channels and regulations and seek food directly from peasant producers." The urban search for clean food reflects broad public awareness of a tradition that harks back to imperial times, in which rural farmers produce two sets of crops and livestock: Set A is grown outside the family courtyard for unrelated customers, while set B is cultivated within the compound gates for its residents (Zhou, Pan, and Dong 2012). This tradition has evolved since the late 1970s, following Chairman Mao's death and the subsequent introduction of market reforms under Deng Xiaoping. Under more open economic conditions, set A has increasingly been produced with pesticides, fertilizers, antibiotics, and other agrochemicals to generate the largest possible yields and revenues. Farmers have defended the legitimacy of this practice as inequalities between rural and urban populations and their livelihoods deepen (Yan 2012). Meanwhile, they continue to follow traditional farming methods for set B, employing organic techniques that prioritize safety and cleanliness for family consumption.

The A-B model, as it has come to be known, is provoking urban consumers to seek out chemical-free fruits, vegetables, and meat directly from producers who are willing to commercialize their B crops rather than reserve them for their families. But their pursuit of food safety is impeded by two problems: First, few small farmers have the technical and labor capacity to scale up production of organics; and second, few city dwellers maintain social connections with small farmers. To bridge these gaps, intermediaries have identified opportunities to generate income for rural producers and, critically, to "build *trust* among people in a social context where trust was abdicated to anonymous government inspectors and tenuous regulations" (Zhang and Qi 2018, 12, emphasis in original; also see Zhao and Wang 2013; Zhang and Qi 2015). These efforts validate the preference of farmers to produce clean B crops while extending the range of recipients to include intermediaries and their urban networks. Resulting initiatives, therefore, seek to strengthen inward community bonds as a basis for building outward bridges, with the broad goal of rebuilding societal trust in Beijing's food system.

The migration of young people from the city's surrounds into its suburbs has provided a foundation for rural–urban connections. The following case study describes how this process has underpinned the development of the Shared Harvest initiative in Beijing's northeastern Shun Yi district. Providing corn, mushrooms, pumpkins, potatoes, poultry, pork, and other produce directly to customers' doors and to retail outlets, the project aims to integrate rural farming traditions with emerging urban preferences. Its leader is agricultural scientist Dr. Shi Yan, who believes that Shared Harvest's success depends, above all, on stimulating a "concentric ring" of trust.

Case Study: Shared Harvest

Established in 2012 as an independent agricultural cooperative, Shared Harvest supplies more than 1,500 Beijing families each week with organic fruits and vegetables. Dr. Shi Yan has overseen the initiative's growth: By 2017, it employed forty staff and expanded its range of fruits and vegetables from an initial base of twenty to eighty (interview, July 31, 2017). When I last visited in 2023, it had 160 staff onsite and, in addition to households, was also supplying private businesses and restaurants with products that now included dried yak meat, organic tea, bottled drinks, and other specialty items. Beijing's supermarkets ship produce from farms across China, but Shared Harvest has developed its brand around reduced food miles, safe organic produce, and relationships between producers and consumers.

Shi studied organic agriculture at Earthrise Farm in Minnesota in 2008, and upon return to Beijing, found that the enthusiasm for the urban farming she witnessed in North America was not widely shared, despite growing consumer concerns over food safety. She reasoned that this was because of limited public knowledge about food system sustainability and set about raising awareness, quickly attracting more than 30,000 followers to her Weibo account. The municipal government granted her fifteen hectares of productive land until 2030 and a methane bio-gasification energy plant to power the farm and produce fertilizer. On the back of this support, Shi founded Shared Harvest with 2,000 chickens, 50 pigs, and a long-term strategy for building customer relations. According to Shi:

> Customers want to see with their own eyes where their food comes from, so we invite them to visit us, and if they wish, to participate in the productive process. They can rent a thirty-square-meter plot and produce their own vegetables. Most

of our customers don't have time to do this, so each week we deliver our produce directly to their doors in Beijing. Our model links the farmers to consumers directly, and in this way we are restoring trust between the two. (Interview, July 23, 2015)

Inviting customers to visit Shared Harvest, meet the farmers, and become involved in production creates relationships that are deeper than those they experience with supermarket chains. Many Beijing retailers sell produce that is labeled as organic, but customers are skeptical about certification; similarly, bruised or damaged fruits and vegetables provoke suspicions about excessive storage times. In contrast, Shared Harvest opens an opportunity for customers to understand the productive process. Visiting in person allows them to experience the reality of organic farming and to understand that a bruised apple or tomato results from movement in transit rather than lengthy storage. As Shi noted in a BBC radio interview, "Right now, consumers care more about the food quality than the price" (Shi 2015). Perceiving an opportunity to build momentum behind emerging interest in safe and sustainable food, Shi established a "New Farmer" training program that by 2020 had inspired the establishment of over 1,000 private community farms across China (Lyu et al. 2020).

When I first visited Shared Harvest in 2015, Shi was planning an international academic conference with Tsinghua University. The event, she reasoned, would attract foreign endorsement, create opportunities for technical support, and raise local awareness about the benefits of organic food. The strategy worked: the conference hosted ninety specialists from the United States, Australia, Canada, the United Kingdom, and Latin America and generated media coverage that helped China's Community Supported Agriculture Network to achieve official registration in 2017. The network's registration gave Shared Harvest, a core member, legal coverage to subsequently host its own meetings and conferences with foreign participants. These extend beyond academic exchange into commercial networking and coordination of policy proposals.

Shared Harvest's quick success caught the eye of policy analysts and scholars such as Wu Wenliang, dean of the School of Resources and Environmental Sciences at China Agricultural University. The initiative, says Wu, "may provide what urbanites need . . . You can trust it because you can always find the producer or farmer and voice any concern. When a farmer is paid in advance, he takes care about the quality. When consumers commit to a membership, there is trust" (quoted in Sun 2013). Shi's mentor, Wen Tiejun,

Figure 3.1
Shared Harvest, Beijing. Photo by the author.

argues that public trust in Shared Harvest derives from its capacity to alleviate pressures arising from Beijing's population boom:

> Never mind the official numbers; Beijing probably now has more than 30 million people living in it. Go back to the Asian Financial Crisis of the 1990s and the Global Financial Crisis twenty years later. Both diminished China's exports and led to labor overcapacity, so the government paid for the new social services by appropriating around thirty percent of rural income. Consequently, millions of people had to pack up and move to the city. But rural migrants can learn to become middle-class entrepreneurs, and Shared Harvest is the kind of small business that can absorb labor, providing opportunities while helping the government by reducing class struggle and street protests. Add to this the respect Shared Harvest pays to regenerating local cultures and you have the basis for long-term cooperation and trust. . . . Traditional culture is full of lessons for the future, and all civilizations have their origin in farming. Take this proverb for example: 民以食为天 (Mín yǐ shí wéi tiān; people regard food as heaven). (Interview, July 19, 2016)

As well as appropriating rural revenues, the national government has deflated the price of fruits and vegetables to discourage reliance on farming and to augment pressure for migration to cities. Shared Harvest's business model, based on prepaid customer subscriptions, responds to this pressure. Farmers enter an occupation that harnesses their skills and beats the prospects of Beijing's precarious construction and informal sectors. In its first year, subscriptions and orders generated just 50,000 RMB (USD $8 thousand), but by 2016, annual income was around USD $1.5 million, and by 2022, surpassed USD $2.5 million. Upscaling mitigated the price fluctuations and unpredictable weather of rural farming and earned the project's workforce greater income and job security than they might expect from low-skilled urban employment.

Shared Harvest's farmers maintain links to their rural families, many of whom have become suppliers of products particular to their climates and soils. The New Farmer training program has facilitated this integration, but the lessons flow in both directions. Among these lessons is that family farmers often prefer not to share their traditional practices and techniques with outsiders. Accommodating this preference is an important step in building trust, as it acknowledges rural producers as valued partners whose contributions are unique and indispensable. Purchasing their food while respecting their privacy validates their adherence to the longstanding A-B tradition. While they may still produce A crops for unknown customers, working with Shared Harvest extends their definition of family outward from traditional courtyards to relatives in the city and the networks they join. Widening the range of recipients of B crops in this way does not erode inward bonds; on the contrary, it strengthens them through the reciprocal benefits (training, income, etc.) introduced by urban relatives. Such exchanges require the backing of a network capable of transferring skills and connecting with markets, and through this process, harnessing inward community bonds as a platform for outward bridges.

When I visited Shared Harvest 2023, its community had grown and its New Farmer program had expanded. I noted some observations in my diary:

Next Generation Education

Over the past twenty years I've seen Beijing's subway system extend outward from two lines to twenty-five, and concrete apartment towers spring up on the city's edges where trees and crops once grew. Incredibly, the two-hour journey northeast from Renmin University to Shared Harvest remains the same: a line 15 train to the last stop at Fengbo, a 23 bus to Liuzhuanghu terminal, and a ten-minute walk to the farm's front gate. I haven't seen Shi Yan for about six years, and she welcomes me together with four others: "Great, now you're here. Let's have lunch together."

The table is set with scrambled eggs, spinach, pork stew, bread, packaged cookies, and pear wine—all produced organically at the farm. The range of

(continued)

Next Generation Education (continued)

products has evidently expanded, and judging by the conversation, so has the community. The time has come, says Shi, for Shared Harvest to start a kindergarten: "In public schools they feed kids instant noodles wrapped in plastic. What kind of lesson is that? The health of our bodies and of our planet seems have been removed from basic education, but we want a nourishing diet to be the foundation for learning." The conversation turns to logistics: Which areas of the farm can be used for teaching, for resting, for preparing food, and so on? Twenty years of studying Chinese, on and off, should have prepared me for discussions like this, but I feel like I'd benefit from studying alongside the children.

Two teenagers enter the room and offer to take me around the farm. They are joined by two others, all of them students from Hebei Agriculture University, living at Shared Harvest for six months to study its marketing and distribution systems. Their timing is good, as Shi must leave now to teach a class for the New Farmer program. This cohort of students, she says, joins her each week online from Tibet. We accompany her down the hallway to a room fitted out with video cameras, monitors, and laptops. A young woman introduces herself in perfect English as "the leader of Shared Harvest's social media team . . . could I steal you for a moment? I'd really appreciate a few words for a TikTok post." Before the camera, I do my best to say something educational, but I fear it may not be sufficiently entertaining.

Research diary, Beijing, September 12, 2023

After her class, Shi went to the farm's collective office, where amid piles of books and papers, she extended her energy and patience to indulge me with an interview. We began by discussing the farm's name:

> Shared Harvest represents a value that is the opposite of industrial agriculture, including genetic modification, in which patents and property generate money and benefits for private individuals and companies, but not for the public. We are the other way around. First you need a philosophy of sharing, and only then will the harvest benefit society . . .
>
> About half of the students in the New Farmer training program are from farming families, and the other half are architects or professionals with no farming experience. The students from farming families usually try to convince their parents and relatives to expand the B crops and diminish the A crops, so what they grow for the family in their courtyard can become commercialized, and what they grow outside the courtyard with chemicals is gradually replaced. We advise them to do this slowly, since they haven't yet established themselves in the organics

market and would take on too much risk if they go too fast. To reduce their risk, we offer to sell their clean B crops for a time through our online store.

So, we are building up a network of producers who form the basis of our CSA [community-supported agriculture] network. The network's hubs are run by our New Farmer graduates. They get to know other farmers in their area and offer them the possibility of joining the network if they can produce organically. We have an organic certification system that we call PGS [Participant Guarantee System]. This is widely trusted, probably more than the official TPS system used in supermarkets, which is often not properly tested and verified by inspectors. . . . By using our system, we build up concentric rings of members who help small producers join the network and compete with the unfair supermarket system. What makes the rings expand is *guanxi*. (Interview, September 12, 2023)

At the epicenter of Shi's "concentric rings" are rural families, whose departed relatives and online students introduce them to Shared Harvest, which in turn furnishes connections with urban customers. Underpinning these relationships, she says, is *guanxi*: a traditional Chinese concept of trust based on family ties and shared experience (c.f. Hearn 2016, 14–16). Research on *guanxi* illuminates its incremental formation through exchanges that prioritize emotional ties over instrumental transactions, its resulting capacity to operate alongside formal rules, and its utility in compensating for inadequate legal process (Armony 2011, 40–41; Carrier 1999; Lo and Otis 2003). It stands to reason, then, that *guanxi* networks are central to the consolidation of community-supported agriculture (CSA) in China, where according to Zhang and Qi (2018, 124), lax enforcement of safety regulations has provoked a bottom-up push for food sovereignty:

To solve the whole society's food safety problem, self-protection must expand from passive forms focused merely on close social connections. . . . The possibility of resolving the food safety crisis in China must emerge from our understanding of the strategic spaces for food sovereignty that can be carved by bottom-up initiatives within the tensions of the failing top-down governance mechanisms.

In a context of inadequate oversight of standards from above, Shared Harvest constitutes a community-led, yet state-sanctioned, initiative. Although the growing CSA network operates independently of government control, municipal support in the form of land and infrastructure constitutes an enabling proviso. This support was critical at the project's outset in 2012, when Shi was faced with an untested market and almost no urban knowledge of organic food and its connections with the traditional knowledge of rural migrants. Her Weibo posts and media interviews helped to generate this collective awareness—a second essential proviso for her work—and connect her with a broader network of like-minded foreign enthusiasts. Since then, the New Farmer program and a dedicated social media team have encouraged broader reflection on the repercussions of evolving urban food systems in China and around the world. Their message resonates with the values of citizens: a rejection of homogenous industrial agriculture and an embrace of organic horticulture to improve personal and environmental health.

To integrate local enthusiasm for organic food with global trends in sustainable production requires the constant defense and renewal of policies to protect peri-urban arable land. As commodity agribusiness encroaches inward and cities expand outward, the conservation of productive green space is an ongoing challenge. The expiration of Shared Harvest's lease in 2030 raises questions about the initiative's future and, as Shi concedes in the chapter's final section, uncertainty about her leadership. The confluence of opportunities and threats she faces is comparable with those unfolding in Rio de Janeiro, where two intermediaries—Julio and Fabiola—must contend not only with shifting policies but also with attempts to bulldoze their project figuratively and literally.

Finding Nutrition in Rio

Prior to Portuguese colonization, the bay where Rio de Janeiro now sits bore the Tupí name *Guanabara*, or "Bay that is like an ocean." Explorer Gaspar de Lemos first encountered the bay on January 1, 1502, leading him to rename it Rio de Janeiro (January's River). The city proper was founded on the bay's western shore in 1565 and became the capital of the Viceroyalty of Brazil in 1763. When the Portuguese royal family and some 10,000 of its courtiers fled Napoleon's invasion of Lisbon in 1807, their arrival in Rio made it one of history's rare cases of a colonial city becoming the capital of its colonizer. Less unique was the decimation of Indigenous peoples in Rio's vicinity—Tupiniquim, Tamoio, and at least five other groups—to realize European visions of society, land, and nutrition. As in São Paulo, local yams, sweet potatoes, and farofa (fried cassava flour) made their way into mainstream diets, but sprawling sugar and coffee plantations soon dominated the agriculture economy.

Sugar relied on the labor of enslaved Africans, 5.8 million of whom were transported to Brazil between 1501 and 1866. Lucrative coffee harvests were propelled by an influx of Portuguese, Italian, and other European migrants from the late nineteenth to mid-twentieth centuries. Rio's population matched the immigration boom, growing from 811,000 in 1900 to three million in 1950, and 11.3 million by 2000 (Lahmeyer Lobo 1978; PopulationStat 2023b). Agriculture continues to stimulate migration to Rio, constituted now largely by rural workers seeking new lives as mechanized soybean plantations take over their provincial towns. Between 2000 and 2023, Greater Rio's population grew from 11.3 to 13.7 million, adding further density to underserved zones.

A *New York Times* exposé clarifies the link between agribusiness expansion and dietary health: "As multinational companies push deeper into the developing world, they are transforming local agriculture, spurring farmers to abandon subsistence crops in favor of cash commodities like sugar cane, corn and soybeans—the building blocks for many industrial food products" (Jacobs and Richtel 2017). The report notes that companies like Coca-Cola, McDonald's, and meat giant JBS donate heavily to members of Brazil's National Congress. The consequences of increasingly processed diets are evident in Rio, where a study of 5,000 children found that obesity rose from 6 percent to 18 percent between 1986 and 2016, and overweight rates from 17 percent to 32 percent (European Society of Cardiology 2016). These figures reflect the Brazilian Ministry of Health's findings that between 2006 and 2016, national obesity levels rose by 60 percent, from 11.8 percent to 18.9 percent of the population (Governo do Brasil 2017). The land use transformations and demographic shifts associated with an increasingly industrialized food system have exacerbated dietary health complications, particularly in vulnerable sectors, while doing little to create jobs for metropolitan residents. It is a familiar predicament in disadvantaged urban suburbs, also evident in Melbourne and São Paulo (chapter 2) and across a range of case studies from the United States curated by Alison Hope Alkon and Julian Agyeman (2011, 13–16).

As Brazil's second most populous city after São Paulo, Rio houses a fifth of its residents in underserved *favela* communities, but hardship has provoked experiments to boost access to employment and food through urban farming (Catcomm 2023). A policy developed by the federal government and implemented by the Rio municipal council has given rise to fruit and vegetable farms within some of Brazil's most disadvantaged zones to improve nutrition and generate income. Socioeconomic inequality does not favor the formation of trusting relationships across racial, class, and geographic lines, but the common ground generated by organic farming has sustained unlikely alliances in Rio (Jordahl 2007; Uslaner and Brown 2005). The next case study shows how a charismatic agronomist and a political activist have brought together state subsidies, grassroots activism, and the patronage of consumers around the city to transform living conditions.

Case Study: Hortas Cariocas

In 2016, Brazil's *favelas* were officially renamed "self-constructed communities." The shift in terminology underscores the sensitivity of economic and racial inequality in Brazilian cities, both of which are plainly evident in the Manguinhos district of central Rio. With 36,000 residents squeezed into 535,000 square meters, the community is a densely populated amalgamation of eleven informal settlements. Taking its name from the mangrove ecosystem (*mangue*) that once predominated, the neighborhood has become one of Rio's most polluted, with high unemployment and an average per capita income of 430 reais (USD $83) per month in 2023, significantly lower than Brazil's official minimum monthly wage of 1,077 reais. Combined with a scarcity of affordable fresh food in the area, these conditions have against all odds stimulated the creation of Latin America's largest community farm.

The Manguinhos project is managed by Hortas Cariocas (Rio Gardens), an initiative established in 2006 within Rio's Municipal Secretary for the Environment (SMAC) to promote social inclusion and improve nutritional health indicators. With the support of a government-funded research center, the Pereira Passos Institute (IPP), Hortas Cariocas conducted a survey of income and opportunity in areas with land available for horticulture. Forty locations were selected to trial the introduction of community fruit and vegetable production, and with funding from the municipal government and UN-Habitat's Rio+Social program, Manguinhos became emblematic of the program.

Having visited the community regularly between 2014 and the onset of the COVID pandemic, I returned in 2023 to implement a project funded by the Australian government's foreign aid program (described in Hearn 2023). Tasked with providing water tanks, hosepipes, and other materials for medicinal plant production, I recorded some impressions in my research diary:

Trouble and Trust in Manguinhos

Our car is the second of three to pull up around 9:30 a.m. next to the busy main road, a couple of blocks outside Manguinhos. Jackets are unzipped and left on seats as the bright sunshine cuts through crisp morning air, ushering in a warm winter day. "*Tudo bem*" (all good) says the Hortas Cariocas driver into his phone and then tells us we have permission to enter. This time feels different. My past visits were about research, but this time it's both theoretical and practical. Ezequiel [a community leader] says they want equipment to grow medicinal plants, so my role has changed: Now I must put words into action.

We thread our way through narrow laneways, a web of electricity cables running overhead in the cool shade of cramped houses. Their walls are a tapestry of uneven orange bricks patched together with rough grey mortar. Many have two floors and seem to be under construction, or as the saying goes, *sem acabamento* [literally "without plaster," meaning unfinished, unregulated, and constantly growing]. Just as I regret leaving my jacket behind, we emerge from the stony shadows into a vibrant green farm with around twenty people collecting beans, onions, and herbs. I don't recognize people I knew before COVID. Like Ezequiel and Julio told me on WhatsApp, there has been a fast turnover here. Wearing a green Hortas Cariocas T-shirt, a woman of perhaps sixty years, Rafaela, confirms their point: "I'm one of the only originals left. I have ten grandchildren who sometimes work with me here, but once they get to a certain age, they lose interest or leave the neighborhood."

Fabiola [the newly appointed director of Hortas Cariocas] joins us and points out a group of unused garden beds—wild grass overgrowing their rectangular concrete walls—that we can use for the medicinal plant project. "This is the area I told you about" she says, "where we can teach people about *plantas ancestrais* [ancestral plants]." In our Zoom meetings over the previous months, we've talked about using medicinal plants to build a sense of trust and connection with Afro-Brazilian and Indigenous history. Standing in this green oasis amidst a team of busy gardeners, it all seems less abstract and more possible.

The rumble of an approaching motorcycle grows loud, bringing me back from my reflections. I notice people walking at a faster pace, and that some have guns tucked into their jeans. Should I be worried? There's no indication of urgency among my companions, but the decision is made to leave. It's an abrupt and sobering exit from a seemingly harmonious place. We arrange to regroup at the SMAC, but not more than three blocks away a police barrier prevents our departure. An officer with a rifle and bullet-proof vest asks me and one other to step out of the car, and then pats us down. "Are you carrying drugs?" I'm reminded that outsiders have long used Manguinhos as source of cocaine and that police violence has further stigmatized its largely Afro-Brazilian community. I'm not sure medicinal plants can do much to restore a sense of trust and connection.

Research diary, Rio de Janeiro, June 7, 2023

Figure 3.2
Harvesting crops in Manguinhos, 2023. Photo by the author.

While some aspects of life in Manguinhos had changed, the underlying dynamics resembled those of my first visit to the community in 2014. On that occasion I was met by Julio César Barros, an agronomist trained at Minas Gerais University prior to taking up a position in urban forestry within Rio City Council and subsequently the directorship of Hortas Cariocas. He described the process of establishing horticulture gardens as a mediation of official priorities and community interests, which by 2016 had exposed the range of forces at play in Manguinhos:

> The municipal government wants to achieve several goals through our project, including reducing social inequality and slowing the illegal occupation of land. But the people in Manguinhos and our other sites are more concerned with generating income, improving their living conditions, and creating lifestyles that provide an alternative to drugs and other illegal jobs. To attract local buy-in we work with neighborhood associations to identify the participants and a leader of the project who should have some horticultural experience and a capacity for bringing people together. In this way the community gains a sense of ownership. Without genuine local interest the project can't proceed. (Interview, June 20, 2016)

The SMAC, whose funding derives from oil sector royalties and municipal infringement fines, incentivizes local participation by paying a monthly salary slightly above the official minimum wage to a selected team of community participants and an additional 30 percent to the team leader (O'Reilly 2014, 46). In return for this subsidy, half of the produce generated by Hortas Cariocas projects must be donated to fresh meals programs for public schools, childcare centers, and families in need. This policy reflects a broader national attempt to link public schools, underserved communities, and families through urban farming (MDAS 2023). The other half of the produce can be commercialized by the teams for sale in local markets. This business model is intended to evolve over time: As production grows, total commercial value (including of the donated produce) should eventually exceed the sum of the SMAC salaries and the commercialized profits. At this point the community can choose to stop receiving the subsidy and commercialize 100 percent of production, leading Julio to describe Hortas Cariocas as "an incubator of small businesses" (quoted in Barbon and Cheibub 2021).

Of the forty gardens established by Hortas Cariocas around greater Rio since 2006, seven have achieved autonomous status, four are in the process of formalizing their autonomy, twenty-one are receiving the state subsidy, and eight have been suspended because of unsatisfactory results. Inaugurated in 2013, Manguinhos is still subsidized and is the largest and most prominent of these, not least because of the obstacles it continues to face. A thematic challenge has been intense competition for territorial control, deepened by the omission of urban agriculture as a category for land allocation in Rio's 2011 Master Plan (*Lei Complementar*).

Like other Rio communities, Manguinhos has been subjected to the government's "pacification" campaign, which has involved violent conflicts between security forces, narcotics traders, and, all too often, innocent residents. In some cases, "pacification" has provided a pretext for unrestrained use of force by police, undermining the efforts of activists to establish urban agriculture programs in Rio's underserved communities (Rekow 2015, 2016). This is the case in Manguinhos, where a confluence of violent confrontation, drug trafficking, and corruption continue to impede the project's efficacy and local trust in authorities.

Shortly after Julio inaugurated the vegetable garden, a municipal official authorized its main planting bed to be bulldozed and converted into a dump for household and industrial waste. Julio learned that narcotics traffickers had pressured the official to take this action because they were concerned the garden would disrupt their hold on the local economy and intensify external monitoring of their activities. Unwilling to confront the traffickers directly, Julio elaborated an alternative strategy. Several other Hortas Cariocas projects were selling their produce into farmers markets patronized by upper-middle-income clients interested in the broader expansion of organic farming and its positive socioeconomic effects. He engaged these customers in conversations about Hortas Cariocas that resembled his discussions with school principals. The encounters I witnessed focused on the need to defend the social and ecological benefits of small farms in the face of export-oriented agribusiness and the displacement of farmers to Manguinhos and other marginal urban zones provoked by expanding soybean plantations. In the face of a changing world order marked by China's growing demand for food, he argued, consumers should think globally and act locally by supporting the Manguinhos farm.

Inspired by these arguments, some farmers market patrons pledged to email or call the office of Rio's mayor at the time, Eduardo Paes, naming Manguinhos as an exemplary project. The mayor's office asked the local official to halt the destruction of the vegetable garden, and although Julio could not discern the influence of the phone calls and emails, he spoke of multiple agendas coalescing behind a common goal:

> People often say that urban agriculture will never take off because there is no agreement about its purpose. Some support it for environmental reasons while others do it because they want organic diets. Some have political motivations while others do it simply to provide food for themselves and earn a living, like in Manguinhos. But when you build relationships and understanding between these groups the sum of the parts is bigger than the pieces. (Interview, November 21, 2015)

The "relationships and understanding" that Julio engendered between distinct groups resulted in a truce between Hortas Cariocas and the Manguinhos official. This informal accord enabled the project to grow to two kilometers beneath the electricity lines running through the district. The project's profile was raised further through a January 2017 visit from the new mayor, Marcelo Crivella, who pledged to emulate it in sites around the city to feed 1,500 public schools (Lucas 2017). During this phase of expansion, I joined Julio on visits to schools that he hoped to recruit into the program. The purpose I served became apparent when he asked me, at each stop, to discuss how global agribusiness to feed foreign cities has precipitated urban food security risks and how community farming and youth education can help mitigate these. As illustrated at the chapter's outset, he often pointed out that China's unprecedented demand for commodity crops from the world's major agricultural exporters, including Brazil and Australia, has intensified rural–urban migration and compromised peri-urban fresh food production. An accompanying journalist noted that my participation in these encounters demonstrated that "we are facing the same global challenge," strengthening Julio's argument that public institutions in Rio must keep pace with foreign counterparts in the search for local solutions. He further emphasized this agenda in the Facebook page he created to attract foreign visitors, showcase recent outcomes and endorsements from participating communities, and generally raise public awareness about the benefits of localized food systems.

The time and energy Julio invested in Manguinhos were evident in his interactions with residents. He was easy to spot socializing at neighborhood bars, filling the atmosphere with his characteristic deep laugh and standing up enthusiastically to tell animated stories. These occasions became more frequent after he broke his leg in a soccer match and dedicated himself to exchanging MP3 files of AC/DC, Led Zeppelin, and his other rock favorites with local friends. On a more solemn occasion, he recounted to me before a small crowd of onlookers and a distressed family how the police had stormed the district in an armored jeep that week, killing one of their young relatives.

The incident provoked conversations about the lack of jobs and economic opportunities in Manguinhos and personal reflection on my own biases. On another occasion, gathered around a white plastic table one cloudy afternoon in 2017, I asked a neighbor his thoughts on alternatives to "pacification" for neutralizing the threat of drug traffickers. After an uncomfortable silence he responded: "Drug traffickers? You're talking about us." Over time it became clear that Julio's involvement in the joys and sorrows of life in Manguinhos entailed a deep empathy for local conditions that went beyond professional calling and was inseparable from his efforts with the project.

The farm's impact became clear during the COVID pandemic, when it produced two tons of fresh food per month for around 800 families (Barbon and Cheibub 2021). Community members I spoke with in 2023 and 2024 attributed this productivity to the temporary suspension of police "pacification" operations by order of the Federal Supreme Tribunal (STF). According one of the gardeners, Berto, he and his neighbors became less anxious about violent conflict between security forces and criminal groups, "so we worked more hours and let the kids spend more time outdoors with us" (interview, June 7, 2023).

Positive results in Manguinhos generated two problems that ultimately threw the project back into crisis. First, conservative mayor Eduardo Paes (reelected in 2021) argued that the farm's success during COVID diminished the need for state support. He consequently terminated a monthly subsidy of 95,000 reais (USD $18,300) that had provided Hortas Cariocas with trucks, logistical support, and three agronomists. When Julio queried this decision, he was informed that the funds were destined for a new project linking existing urban farms in Madureira, nine miles west of Manguinhos, which would enable Paes to claim oversight of the largest urban farm in the world (Boas Noticias 2022). With his funding cut, Julio left Hortas Cariocas and transferred leadership in Manguinhos to Ezequiel Dias, who had been a committed team member since 2013. Ezequiel tried to raise the project's public profile by participating in media interviews, telling a journalist in 2022 that if it were not for the garden, "today I could be selling drugs, or dead, or in prison" (Boas Noticias 2022).

Ezequiel's reflection pinpoints the second problem: After the suspension of "pacification" operations in 2021, narcotics traffickers reasserted control. The government's all-or-nothing approach to security—reminiscent of Mexico's "*mano dura*" under Felipe Calderón (2006–2012)—left no room for community-oriented policing. Julio lamented that by early 2023, the withdrawal of state subsidies, the related lack of investment in socially engaged policing, and the consequent reemergence of violent conflict between security forces and gang leaders had "paralyzed" the Manguinhos farm (interview, June 8, 2023). Neighbors had little faith in the mayor's supposed commitment to funding Hortas Cariocas, and several offered to join Julio in his new project: a small private business providing equipment and technical support to community gardens across the state of Rio.

The contours of change in Manguinhos continue to follow political developments, particularly the 2023 reelection of the Worker's Party (PT) candidate, Luiz Inácio "Lula" da Silva. A month after Lula's January inauguration as president, Tainá de Paula, a prominent PT supporter and outspoken activist on issues facing women of color, was appointed as Rio's new secretary of environment. Tainá's political alignment with Lula strengthened her position vis-à-vis the conservative mayor, Eduardo Paes, who she successfully pressured to reinstate funding for Hortas Cariocas. The truce enabled Tainá to recruit Fabiola Corrêa, a school teacher from the Rio *favela* community of Santa Marta, as the new Hortas Cariocas director. Following the appointments of Tainá and Fabiola into these so-called *postos de confiança* (positions of trust), the Manguinhos farm became explicitly connected with a progressive embrace of social inclusion and ethnic pride. Fabiola shared her reflections on building consensus around this agenda:

> Growing up in Santa Marta and working as a teacher there gave me the skills to negotiate. I've always known difficulties so maybe I've become empathetic to what different people want. . . . I've realized that the best way to achieve stability is to create public policies, and I see this as one of my main responsibilities. So, I'm trying to create public policies that work for people in communities like Manguinhos. . . . It's not easy but I'm always ready to learn what motivates people to collaborate. In Manguinhos one of the things that brings people together is recognizing the traditional knowledge of Afro-descendants, for example, through *plantas ancestrais* [ancestral plants]. (Interview, June 9, 2023)

Fabiola's efforts to embed cultural inclusion in the Manguinhos farm offered an unexpected overlap with my project to support the cultivation of medicinal plants. From 2021 to 2023, I had worked on a similar initiative in Havana with the Afro-Cuban community association Ifá Iranlowo (SOM 2023), whose positive results were an important factor in securing funding for Rio. Working in Manguinhos on the ancestral plants project with Fabiola and Samuel (introduced at the book's outset) provoked reflection on the social and political conditions that had formed over the preceding decade.

Prior to Fabiola, Julio's efforts to bridge the interests of distinct interest groups had been tested by extreme socioeconomic inequality, opposition from vested interests, and corruption of local officials. The formation and erosion of collective trust in this environment demonstrate how policy settings, public familiarity with food traditions, and job security are key provisos for effective mediation. A municipal commitment to supplying public schools with fresh food constituted an enabling policy framework for Julio. This was strengthened by the city government's interest in harnessing residents' horticulture skills to stimulate the emergence of small private businesses. Resulting subsidies underpinned the Manguinhos farm's growth and impact from its establishment 2013, but the Paes administration's withdrawal of this support in 2021 took a heavy toll. Despite Julio's protests, the project was subordinated to the mayor's pledge to create the world's largest urban farm in the nearby suburb of Madureira. Lula's 2023 election and its ripple effects once again turned the tide in Manguinhos with the appointment of Tainá de Paula as secretary of environment. Hortas Cariocas, now led by Fabiola, secured fresh funding to resume its activities despite the ongoing troubles described in the diary excerpt.

The second proviso, public familiarity with community farming, was fortified by the school meals policy and evident—as in chapter 2's analysis of São Paulo—in collective support for traditional horticulture as an alternative to industrial agriculture and processed food. This background knowledge became critical when the local government attempted convert the Manguinhos garden into a dumping ground. Julio's conversations at schools and farmers markets in wealthier Rio suburbs harnessed public sensibilities about urban agriculture's historical benefits, generating political pressure to halt the farm's destruction. Fabiola's promotion of "ancestral plants" also drew on background knowledge, initially to strengthen the community's inward bonds of ethnic heritage and identity. Her progressive approach centered on rebuilding trust between municipal institutions and a traumatized community.

The shifting priorities of successive mayors brought uncertainty to Manguinhos. Julio's professional status mirrored the farm's development: Both persisted through conflict, narrowly survived sabotage, and suffered debilitating budget cuts. Lula's return to power in 2023 ushered in a more favorable institutional context for Hortas Cariocas, yet it remains to be seen whether Fabiola can build public policies that protect the stability of the initiative and its leadership. Job security is a key proviso for the sustained effectiveness of food system intermediaries, and yet in Manguinhos and elsewhere, it has eluded most. Like Jack's fabled beanstalk, intermediaries are valuable for the connections they provide but vulnerable to the axe. The chapter's final section suggests that reliance on individuals under these conditions poses risks to multistakeholder projects. Discussions with Shi in Beijing, and with Julio and Fabiola in Rio, reveal the need to cultivate approaches to leadership that are more democratic and less personalized.

Valuable Is Vulnerable

A rare ability to build collective trust underpins the importance of Shi to Shared Harvest, Julio and Fabiola to Hortas Cariocas, and other intermediaries to their projects. But individual esteem does not make intermediaries immune to personal problems and political pressures, which can undermine the endeavors they lead. Aggravated by a lack of professional status, their precarity evokes the maxim of the Dutch poet Lucebert: "Everything valuable is vulnerable."

When I asked Shi, Julio, and Fabiola what would become of their projects were they to fall ill or become indisposed, they acknowledged the problem and offered some reflections. Shi was confident that Shared Harvest's longevity was secured by the "power of tradition" because, she said, emerging urban appreciation of ancient connections to land and rural *guanxi* relations, coupled with the government's longstanding commitment to food security, supersedes the circumstances of any individual leader (interview, July 31, 2017). She conceded, though, that she or a new director would need to build and broker fresh relationships to extend Shared Harvest's territorial lease beyond 2030. Undertaking this work would require somebody with a rare blend of managerial skills and cultural sensitivity.

Increased delivery orders during Beijing's protracted COVID lockdowns and stay-at-home mandates strengthened Shared Harvest's prospects. By 2023, the number of employees had doubled to 160, injecting new labor and knowledge. Relatives of the farm's staff and students continued to follow the A-B model, but they were shifting their production toward the organic B crops as Shi's "concentric rings" emanated outward from villages and cities around China. She was confident that these expanding rings, driven by common experiences of urbanization, concerns about industrial agriculture, and reliance on *guanxi* relations, were gradually restoring societal trust in the food system. The pace of growth, she said, was unexpected and had prompted her to consider new approaches to Shared Harvest's leadership:

> It was not my intention for Shared Harvest to get this big, so we may divide into separate units. The New Farmer program may soon be led by a former student, Lin. She's an energetic and inspired young woman born in 1995, and she could make it more independent from me. This is about collective leadership rather than the focus being on me. . . . I try to remember something I learned from Professor Wen Tiejun: "There are no leaders; just volunteers." Our CSA model works like that. We pay a lot of attention to educating our members so that they can take charge, rather than being told by someone what to do, but this is not simple because most people are uncomfortable being leaders. I know I won't always be the leader of Shared Harvest because nothing should last forever. This is nature . . .
>
> I used to think it was a problem that so many students left after two years to set up their own farms. But then I realized that inspiring them is our gift, because it takes great courage to create your own farm or to convince your family to change its approach. Now they are following their own dreams and not mine . . . So long as people continue to be inspired by clean and organic food, and building communities around it, then our values will continue to exist—with or without me. (Interview, September 12, 2023)

Despite the challenge of finding members comfortable with leadership, Shi was confident that Shared Harvest would continue to inspire courage among young farmers.

Meanwhile, in Rio, Julio and Fabiola pointed out that Hortas Cariocas continued to endure extreme volatility. In response to my question about the risks of individual leadership, in 2017, Julio recounted an episode that removed him from his post amid tensions with a competitor for school meals contracts. His competitor, he said, was a prominent food distributor who convinced the mayor's office that less perishable root crops, such as the manioc and sweet potatoes that he could provide, would constitute more stable and cost-effective school meals. Julio was ordered to take "forced vacation," and his part-time assistant, Aline, was left to single-handedly manage Hortas Cariocas and its relations with community farms around Rio. Six months after Julio was stood down, the official who had ordered it was suspended on corruption charges relating to the *Lava Jato* (car wash) scandal. Hortas Cariocas narrowly survived the episode, and when Julio returned to work, he lobbied for a budget to employ additional staff. He reasoned that broader, more collective leadership was the best way to mitigate against possible future attempts to sideline him or, equally grave, another soccer accident.

Following the mayor's budget cuts in 2021, Julio left Hortas Cariocas definitively. He took a job with the state government supporting fresh food production for eighty public schools, but this too was short lived. When I visited in 2023, he was setting up a private gardening service in Cabo Frio, east of Rio, catering to small businesses and families. Although the venture also involved risks, he noted that these arose from his own decisions rather than the government's shifting policies. Hopefully, he said with a laugh, Ezequiel might "escape from the *favela*" to join his new initiative, since "I've learned the hard way not to lead by myself." He was confident in the personal abilities of the new director, Fabiola, but less optimistic that she would receive adequate institutional backing. Fabiola had evidently reflected on the intersection of personal leadership and institutional support when she told me, above, that "I've realized that the best way to achieve stability is to create public policies." Later she elaborated on the personal implications of this point:

> Remember, Hortas Cariocas is not an individual thing. It's an institution and it's backed by political support. One thing I think we could do better and work on is training and *capacitaçoe* [capacity building] so that there are people there ready to go if necessary. I intend to set up a training program, *com certeza* [without a doubt]. Right now, if something were to happen to me or I couldn't keep working, we need to have people who can take over. (Interview, June 9, 2023)

Fabiola, Julio, and Shi each noted the challenge of identifying capable and courageous people who might step in for them should the need arise. This may reflect the post-COVID shortage of candidates for available jobs, evident in historically low unemployment rates around the world, but the cause may also lie in the nature of the work itself. Leaders in most companies and institutions have clearly defined responsibilities, contracts, and paychecks, but this is not so for most intermediaries. Brokering informal accords on the fringes of official enterprises and ministries, often with little job security, does not attract an abundance of talented leaders, magnifying the vulnerability of collective projects and impeding the diversification of their management.

The three intermediaries acknowledged the need for less personalized leadership. Shi's expanding "concentric rings" of CSA practitioners, driven by *guanxi* between rural farmers, their uprooted urban relatives, and students of the New Farmer program, would hopefully inspire the appearance of more rising stars like Lin. And Fabiola's intention to establish a *capacitaçoe* (capacity-building) program would increase the pool of potential Hortas Cariocas leaders as the initiative reaches beyond its social and geographic origins to engage a wider public. Emerging from the transformation of community bonds into societal bridges, opportunities for diversification in both organizations are growing, favoring their continuity through time and space. The projects suggest that food system intermediaries function to some extent like beanstalks, and as every farmer knows, multiple beanstalks make harvests less vulnerable to hostile environments.

Rising public interest in environmental sustainability, food safety, and connection with the productive process bodes well for Shard Harvest and *Hortas Cariocas*. This awareness has grown as the consequences of commodity agriculture—deforestation, contamination of waterways with pesticides, loss of peri-urban family farms, unsustainable rural–urban migration, and so on—become more evident. Often associated with these consequences is a geopolitical narrative about dependency on Chinese demand and the implications for national sovereignty. When intermediaries join the dots between food system sustainability, environmental stewardship, and China's rise, they construct a narrative—a *bricolage*—that resonates among local, national, and foreign actors.

The next chapter explores how unifying narratives have also shaped Cuba's development, from a leading industrial sugar producer under colonial rule to a globally celebrated pioneer of urban farming. Case studies show how two intermediaries articulate historically loaded accounts of food sovereignty to forge consensus among local officials, community members, and foreign patrons. Securing their own livelihoods in the process, both try to balance collective goals with personal interests and sometimes rigid political constraints. The resulting tensions reveal a paradox: Cuban food sovereignty influences foreign conditions, while foreign influence conditions Cuban food sovereignty.

4 Reimagining Food Sovereignty in Cuba

At an organic fruit and vegetable garden in the south Havana suburb of Arroyo Naranjo, Elena Casas says she has been cheated. "How can I compete when my neighbors use fertilizers, pesticides, and contaminated water?" She gestures to the gray-green river that snakes through her property and the district's other small farms, its surface reflecting an oily rainbow dotted with plastic bags. Most of these farms, she says, work with a distributor who collects their tomatoes, carrots, and lettuces and brands them as "*orgánico*" alongside hers in the local market. Inaugurated six months prior, in early 2018, to support independent producers and trial the sale of premium organic produce in Cuba's emerging private sector, Elena's neighborhood market features above its entrance a painting of Fidel Castro with the words "*Soberanía alimentaria para siempre*" (food sovereignty forever). Without an organics certification system, the intensification of market competition has led many Cuban food producers to overlook concerns about health, ecology, and patriotism (Nelson et al. 2009, 233).

The following week, at a conference titled "Toward a Culture of Nature" hosted by Cuba's esteemed environmental organization the Antonio Núñez Jiménez Foundation (FANJ), a panel on food sovereignty is underway. A Ministry of Agriculture official squints through the spotlight while raising her finger at the audience: "Ensuring a sustainable and healthy supply for Cubans is more important than commodifying our production. . . . During the Special Period our tradition [*costumbre*] of food sovereignty ensured our survival and solidarity with the world." Her statement invokes the acute hardships and innovations of the 1990s, when the collapse of the Soviet Union induced a 40 percent decline of national GDP and forced Cubans to feed themselves through neighborhood horticulture supported by international NGOs.

Simmering below the surface of this narrative is the paradox of food sovereignty sustained by foreign solidarity, which has manifested since the early 2000s in growing Chinese agribusiness investments and in the proliferation of "study tours" tailored to US universities.

A more visceral tension has presented itself over the past two decades, during the cumulative four years I have lived with Cuban families in Havana and Santiago de Cuba. Visits to vegetable and fruit farms, whether small demonstration gardens in urban schoolyards or large suburban *organopónico* facilities, typically evoke proud assertions from their keepers that food sovereignty is a national value. Come dinner time, though, tomatoes, lettuce, spinach, and other fresh produce make only occasional appearances alongside beans, rice, and (sometimes) chicken or beef. Moreover, although locally grown fresh food is widely available and affordable in neighborhood markets, many Cubans prefer to spend their *pesos* on the fatty processed options that have appeared in the privately operated restaurants and cafés that have proliferated since the 2011 economic reform "guidelines" (*Lineamientos*). Extracting lettuce leaves from a cheeseburger at one such place in Old Havana in 2024, with a can of cola on the side, my friend Miki stated a matter of fact: "I don't eat vegetables." His distaste for fresh ingredients recalls the observation of ethnographers that ingrained memories of scarcity have contributed to an enduring preference for *comida fuerte*, or "strong food," over ostensibly weaker fresh alternatives (Garth 2020; Moon et al. 2022, 72; Pertierra 2012, 50; Rosendahl 1997, 36, 49–50). As discussed later, Cuban food sovereignty policies published in 2020 and 2021 emphasize nutritional health and cultural diversity but stand at odds with historically conditioned preferences and the choices offered by an expanding private sector.

Food sovereignty appears in different guises in these scenarios, serving purposes that generate their own contradictions. The term is presented in each as an unchanging value, yet each demonstrates that values have changed. The divergence has widened through state programs to generate revenue through "agroecotourism" projects like those analyzed by Krystyn Moon et al.: "The Cuban government has spotlighted the island's communal agricultural processes, organic farming techniques and small-scale restaurant ownership to tap international interests in sustainability and attract more sightseers" (2021, 11). These efforts have opened opportunities for universities and international solidarity networks to provide their members with a first-hand look at Cuban food sovereignty, and to document the insights it provides other nations and communities for improving their food ecologies, agriculture policies, and daily diets (Cuba Platform 2020; Gürcan 2014; Reardon and Pérez 2010, 909; Wright and Morris 2015). In this chapter, I will suggest through two case studies that such encounters bring distinct global and local understandings of food sovereignty into contact and that intermediaries have emerged to forge consensus.

The two Cuban intermediaries introduced later carefully position food sovereignty in history to reinforce internal bonds within distinct groups, and then harness these bonds to build bridges between them. Temporality is key to this endeavor: In the first case, a static portrayal of food sovereignty as a patriotic tradition generates an international network focused on one-off visits and quick results; in the second, a dynamic portrayal of the concept as an evolving tradition produces a network seeking long-term engagement and policy reforms. This difference suggests a dialectical flow, in which food sovereignty's portrayal harnesses public sensibilities, while public sensibilities guide its historical portrayal. As in the book's previous chapters, food system intermediaries first and foremost influence *why* networks form and, consequent to this influence, shape *how* they do so.

The intermediaries introduced thus far have shown that the articulation of broadly familiar traditions can strengthen community bonds as a prelude to building bridges between groups. This chapter finds that intermediaries can more effectively harness inward bonds to foster outward bridges when collective affinities are temporally associated with emerging priorities. Their capacity to shape this process in the Cuban context is enhanced by food sovereignty's thematic reverberation through national history. Nineteenth-century popular struggles to end Spanish rule dismantled the colonial sugar economy but ushered in an era of dependency on US consumers; the 1959 Cuban revolution threw off this dependency but provoked the US trade embargo and renewed reliance on sugar exports to the Soviet Union; and post-Soviet austerity since the 1990s opened Cuba to Chinese investment in the sugar industry and to foreign tourism, with food sovereignty emerging as a draw for study tours. Aligning with the push for agroecotourism, the Cuban government's recent use of the term indicates an attempt to steer citizens away from unhealthy food choices, but such attempts must contend with ingrained preferences for *comida fuerte* that—as Miki shows—are increasingly unhealthy.

Upon this rich cultural undersoil, the two Cuban intermediaries, Luis and Reinaldo, refashion food sovereignty in a *bricolage* that demonstrates its emotive appeal and capacity to shape relationships across borders. Navigating shifting policies and professional pressures, their work demonstrates that history matters: How intermediaries portray the cultural past orients the purpose of their projects and the character of their networks. Through this process, Cuba continues—as it has since the 1960s—to influence global understandings of food sovereignty.

Sovereign Dependence

Cuban experiences of agricultural dependency run deep. Spanish colonial experiments with sugar farming began in the early sixteenth century, ramping up after the 1791 Haitian revolution expelled the French colony's plantation owners. At the invitation of the Spanish crown, they relocated to Cuba and transformed their new lands into the world's leading sugar producer and the Caribbean's largest importer of enslaved Africans. European demand driven by the Industrial Revolution entrenched Cuba's dependence on sugar (Mintz 1985), provoking a revolutionary grievance famously declared by José Martí: "A people commits suicide when it entrusts its subsistence to a single fruit" (1963, 21). Despite Martí's warning, production increased in the early twentieth century as US investors wrested control of the island's plantations from Spanish administrators. Generating more than five million tons annually (one-fifth of the world's total) by 1925, Cuba remained the world's largest sugar producer until the 1960s, with 90 percent of harvests bound for the United States (Pollitt 2004).

The revolution led by Fidel Castro in 1959 ended Cuba's dependence on the United States, but the Soviet Union became the island's new big brother and established yet another foreign dependency, this time to feed Eastern Europe's escalating sugar addiction. As the Cold War set in, Cuba's alliance with the Soviet Bloc permeated the discourse of its ministers and officials. The director of the National Institute for Agrarian Reform, Antonio Núñez Jiménez, likened his government's determination to increase agricultural productivity with US and Soviet resolve to visit the moon. "Our voyage to the moon," as Núñez Jiménez described it, is explained by Reinaldo Funes Monzote (one of the two intermediaries featured later) as a declaration of sovereignty: "The Cuban voyage to the moon can be interpreted as the ultimate conquest of sovereignty and self-governance, which put an end to the traditional economic and political dependency on the neighbor to the North throughout a long 'love-hate' relationship with that country" (2019, 466).

While Cuba's turbulent history of Soviet sugar bartering, supply quotas, missed deadlines, and mounting debt are well documented (e.g., Funes Monzote 2019; Tsokhas 1980), China's role in Cuban agriculture is less well known. From the outset of the revolution, Cuba forged a relationship with China premised on the exchange of sugar for political, military, and technical support. On the September 2, 1960, Castro declared that Cuba would sever ties with Taiwan; this was done on September 28, making Cuba the first Latin American country to establish diplomatic relations with the People's Republic of China. In the first half of the 1960s, China provided economic and military aid to Cuba, helping it to repel of the Bay of Pigs invasion (People's Republic of China Ministry of Foreign Affairs 1987, 365–366). In return, Cuba supported China's bid to restore its status in the United Nations and agreed to provide sugar for its growing population. Bilateral trade was initiated and regulated through a noncash system of annual quotas, codified in 1960 in the first five-year trade and payment agreement. The agreement was eminently political, with Premier Zhou Enlai proposing that China would buy one million tons of sugar from Cuba (at the inflated price of USD $100 million) and pay for it with industrial equipment. Ernesto "Che" Guevara recognized the added value of Zhou's offer: "This [proposal] is not tenable from the economic point of view, but we raised it from the political point of view" (Ministry of Foreign Affairs of Cuba 1960, 5).

Despite political tensions provoked by the "Sino-Soviet split" over the leadership of the international Communist movement, Cuba continued to barter its sugar for Chinese manufactured products. Castro's support for China through the 1989 Tiananmen crisis, and Chinese president Jiang Zemin's promise to "save Cuba's revolutionary project" after the Soviet collapse, demonstrated the depth of mutual commitment (Hearn 2016, 35). Although Chinese support at this time underpinned Castro's political survival, the nature of bilateral relations began to change in the 1990s as market liberalization expanded around the world. China terminated the trade and payment agreement and began advising Cuba to pursue economic reforms, ushering in what a Chinese newspaper called a "period of completely new and steady development" (*People's Daily* 1996). The new approach implicitly aimed to diminish Cuba's reliance on China by redefining the boundary between dependency and sovereignty, not least in agriculture.

The general secretary of the Chinese Communist Party, Jiang Zemin, visited Cuba in 2001, pledging to support the production of sugar for sale to the Chinese state, but also rice, corn, and other staples for Cuban consumption. Secured during President Hu Jintao's 2004 and 2008 visits to Havana, Chinese tractors, irrigation, storage facilities, and other agricultural inputs have since appeared across the island. To provide electricity for the Jesus Rabi sugar mill in Matanzas and its surrounding population, China Eximbank has financed a biomass power plant that consumes residue from the mill. Construction, engineering, and initial operations were managed by Shanghai Electric, whose 325 technicians worked alongside 250 Cuban counterparts. The plant is intended as the first of eighteen planned by 2030 to operate with sugar mills across the country (Pérez Sanchez 2020).

Chinese agriculture investments will not return Cuba to its prior status as a global sugar powerhouse, but they are breathing new life into the industry after years of decline. Sino-Cuban annual trade grew rapidly between 2000 and 2008, when it peaked at USD $2.3 billion, and has since declined to USD $862 million in 2023 (figure 4.1). The exception to this general decline was Cuba's sugar exports to China, whose value grew from USD $63 million in 2000 to USD $170 million in 2020, only to collapse to USD $13 million in 2023 in the wake of COVID and amid an intensifying fuel shortage (UN-COMTRADE 2024). Indeed, sugar's decline to 350 thousand tons in 2022–2023 (from 473 thousand in 2021–2022) marks the lowest harvest in more than a century (Brizuela 2024).

Declining trade, suppressed further by the COVID-19 pandemic, substantiates reports from Cuban diplomats that their Chinese counterparts want to avoid deepening dependencies. As a former Cuban embassy official in Beijing, Omar Pereira Hernández, told me, "We can no longer knock on China's door and ask for favors the way we used to. Political trust is still important in our relationship, but now it's a matter of pragmatism, because the Chinese insist that the numbers have to add up" (interview, May 28, 2012). Cuba's former ambassador to China, Mauro García Triana, described the situation to me this way: "The Chinese are very clear about one thing: they're not going to be benefactors for Cuba like the Soviets were. I was once told in no uncertain terms by a Chinese diplomat: 'Our relations with Cuba have to be mutually beneficial or they will not work'" (interview, January 11, 2007).

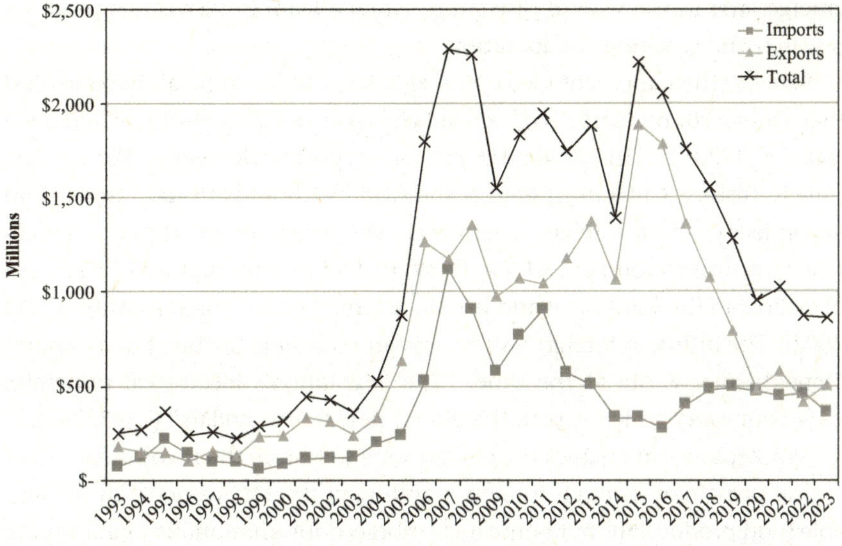

Figure 4.1
Trade with Cuba as reported by China, 1993–2023. Source: UN-COMTRADE (United Nations Commodity Trade Statistics Database) 2024.

Bilateral commercial agreements have been accompanied by Chinese advice about stimulating food production for Cuban households rather than for export, a goal born out in cooperation through China's Belt and Road Initiative (*Global Times* 2021). Among the projects underway is the construction of a facility in Pinar del Rio province by the Chinese enterprise Muyang, which aims to process thirty-seven tons of rice per day for consumption in Cuba (Hernández Cáceres 2020). Unlike investments in the sugar industry, such projects aim to build the island's domestic food production capacity, supporting the Cuban government's food sovereignty narrative even as the influence of Chinese firms on the island deepens. Criticism of this influence emanates mainly from conservative US think tanks and commentators, who allege, for instance, that cooperation in sugar and rice production, oil refining, and telecommunications constitute attempts "to prop up the Cuban regime" that ultimately put China "on the wrong side of history" (Lazarus and Ellis 2021). From this perspective, Xi Jinping's vision of a "global community" and "new era" of geopolitical alliances may provoke countermeasures from the United States aimed at slowing Sino-Cuban cooperation (Romero 2023, 297). Concerns about China's impact on Cuba's food system emerge later in the second case study, when a team of US visitors proposes research and teaching collaboration.

Beneath their apparent ideological alliance, Chinese officials have advised their Cuban counterparts to relax regulations on foreign investment and trade (Mao et al. 2011). This advice reflects a sober post-Soviet reality: For the first time in history, Cuba must engage the world without patronage from—and dependency on—a foreign superpower. This predicament explains Cuba's reluctant turn to tourism, which, by 2001, had quickly captured 11 percent of visitors to the Caribbean and has since remained its largest industry (CTO 2002). The influx of foreign visitors was an economic lifeline, but as Andrés Oppenheimer wrote at the time, "The Revolution's ideological principles were bent every which way in the bid for new tourist dollars" (1992, 286).

Agroecotourism enables the Cuban state to restore the revolution's ideological principles through foreign branding of its achievements in sustainable food production. But while it is embraced internationally as a unifying theme, on the ground food sovereignty is articulated to serve a wide range of agendas, including those of entrepreneurial small farmers seeking greater independence from the state. The next section finds that food sovereignty's ambiguity does not diminish its efficacy; on the contrary, it enables intermediaries to temporalize the concept as a Cuban tradition in ways that suit their projects and strengthen their networks.

Food Sovereignty as Mediated Message

Advocated by politicians, analyzed by scholars, and presented to customers at markets like Elena's, food sovereignty permeates Cuba's recent history. Fidel Castro did not initially use the term, but one of the first actions taken by his revolutionary government in 1959 was to transfer 40 percent (later 70 percent) of arable land from private to state ownership to "satisfy the nutritional needs of the population" (Schultz 2012, 123). Disgruntled landholders lobbied in Washington, DC, for the US trade embargo on Cuba, whose imposition in 1961 inaugurated an era of dependency on Soviet oil, fertilizers, and pesticides that by the late 1980s had exhausted soils and compromised productivity (Machín Sosa et al. 2010, 41).

The collapse of the Soviet Union forced Cuba to reckon with the legacies of dependency just as farmers across the Global South coalesced in the *Vía Campesina* movement to advocate food sovereignty. The term was formulated as a bottom-up alternative to "food security," which the World Trade Organization's 1993 Agreement on Agriculture proposed as a framework for corporate mass production of grains and other staples. By contrast to food security, food sovereignty emphasized social and ecological wellbeing to address "the gap between the levels of policy and practice, national and global processes and local communities" (MacRae 2016, 229, also see Perfecto et al. 2009, 209–212). Food sovereignty was presented by *Vía Campesina* at the 1996 World Food Summit and affirmed by Fidel Castro (2001) at the World Forum on Food Sovereignty in Havana as "a new term that encompasses a series of new concepts," such as multilateral rejection of proliferating agricultural patents. Food sovereignty was formally defined in the 2008 World Agriculture Report as "the right of peoples and sovereign states to democratically determine their own agricultural and food policies" (IAASTD 2009, 10).

State-led food production for public rather than corporate benefit is a widely appealing antisystemic principle, but it belies food sovereignty's multiple valences. These manifest on the ground in advocacy for causes including government subsidies for small farmers, greater autonomy from the state among entrepreneurial producers, conservation of forests and wetlands, and even private acquisition of nature reserves. The concept can therefore be described as a "civilizational movement" whose "salience is universal, but with distinctive local meaning" (McMichael 2014, 938, 947). Consequently, activist Raj Patel concludes that "Food sovereignty is, if anything, over defined. There are so many versions of the concept, it is hard to know exactly what it means" (2009, 663). Rather than trying to define food sovereignty, some researchers have embraced it as a "multidimensional concept" whose enigma is "a virtue, a strength and a reflection and acknowledgment of on-the-ground diversity" (Edelman et al. 2014, 914, 924).

Appreciating the multiple agendas of food sovereignty movements shifts emphasis away from international policy advocacy and toward the daily sharing of knowledge and resources. Such lived experiences of reciprocity have consolidated in Latin America's *campesino-a-campesino* (farmer-to-farmer) movement, in which Cuba has been influential. By 2001, within a decade of the movement's introduction on the island, it had attracted more than 100,000 Cuban smallholders—twice as fast as in Mexico and Central America (Holt-Giménez 2006, 2009). The integration of this process with government priorities and international movements (including *Vía Campesina*) has been documented and supported by Cuban agronomist Fernando Funes Sr. and his son Fernando (e.g., Funes et al. 2002; Funes Monzote 2008), whose scholarship and activism have informed food sovereignty research more broadly. Drawing on their work, Eric Holt-Giménez observes that Cuban policies since the Special Period have enabled farmers to access to land, credit, and markets, and that state subsidized farmer-to-farmer exchanges are a critical aspect of international food sovereignty (2009; Holt-Giménez and Shattuck 2011). Cuba, he argues, demonstrates a "will to transform" that has generated a "breathtakingly successful urban agriculture program" (Holt-Giménez et al. 2009, 158). The second of the following case studies explores an example of a successful farm operated by the Funes family, whose mediation of distinct foreign and local priorities continues to influence international perceptions of food sovereignty.

Food sovereignty's rhetorical currency in Cuba shows how understandings of the term have been conditioned by the challenges of the Special Period and changes in government policies, international agendas, and dietary preferences. Anticipating the Soviet collapse, a prominent Chinese-Cuban military general, Moisés Sio Wong, began installing small organic farms (*organopónicos*) within armed forces facilities in 1987 (Koont 2009, 45). The foresight of these experiments became evident when Cuba's diesel imports fell by 47 percent, petrol by 75 percent, agrochemicals by 80 percent, and food by 50 percent (Wright 2009). In the early 1990s, Havana residents were consuming only 20 percent of the United Nations Food and Agriculture Organization's recommended vegetable intake (Jatar-Haussmann 1999, xv). At this critical point, the Cuban government invoked food sovereignty as a guiding principle to roll out the *organopónico* program nationally and stimulate domestic production (Chan and Freyre Roach 2012; Schultz 2012, 119). Portrayed in Faith Morgan's (2006) film *The Power of Community*, these initiatives blurred the boundaries between rural farms and available land within cities, where "supporting urban agriculture was one of the government's measures to achieve food sovereignty" (Leitgeb et al. 2016, 423).

The creation of the Department of Urban Agriculture and its 1997 expansion into the National Program of Urban Agriculture enabled centralized policies to encourage chemical-free food production in unused city lots, residential balconies, and backyards (Premat 2009). Reinaldo Funes Monzote explains that this approach departed from years of reliance on commodities:

> The new focus on small scale production was completely different to agriculture policies that for decades, actually centuries, had focused on sugar. The 1970s and 1980s also saw elevated production of eggs, milk, and chicken, using technology mainly from Western countries. But that technology was energy intensive, so when the Special Period hit, the state had to shift to vegetables, and people had to accept it. It was a difficult transition, but it was a matter of survival. (Interview, September 17, 2024)

Prior to the Special Period, public interest in urban farming had been minimal, correlating with decades of indifference and even disdain toward uncooked vegetables in Cuban diets (Benjamin et al. 1989, 103–107; Lowry 1950, 209; Wright 2009). This attitude was likely strengthened by the association of subsistence farming with poverty, alongside government prohibition of food crop cultivation in cities (Altieri et al. 1999; Moon et al. 2021, 74, 2022, 69). During the Special Period, though, localized cultivation of organic fruits, vegetables, poultry, and pork softened the impact of the post-Soviet crisis and drew praise and technical support from foreign advocates of agroecology (Funes et al. 2002). Provoked by necessity more than health or ecological concerns, Cuba's policies did not attempt to address excessive consumption of sugar, whose ready availability through the centuries had made it central to traditional recipes (Wright 2009; Villapol 1992, 119). Then Vice President Raúl Castro affirmed his brother Fidel's tendency to evaluate outcomes in terms of quantity while leaving aside food's cultural or emotional associations: "Today, the principal political, military and ideological task for all revolutionary Cubans is . . . to guarantee food for the people" (quoted in Schultz 2012, 126, also see Pertierra 2012, 46). The "Greening of the Revolution," as Rosset and Benjamin (1994) described it, underpinned Cuba's food security, national sovereignty, and probably the survival of its political system.

The hardships of the Special Period provoked offers of assistance from foreign NGOs, which in some cases were invited to embed their staff in Cuban institutions. The state authorized the creation of domestic NGOs—with varying degrees of autonomy—to partner with foreign counterparts and limit their influence by insisting on national sovereignty as a guiding principle of development cooperation. By the end of the 1990s, the Cuban United Nations Association (2000, 9) reported over two thousand functioning national–foreign NGO collaborations (also see Gunn 1995).

The state's insistence on mediating the interactions of community projects with external collaborators produced tensions with donors, who had to accept that "building civil society" in Cuba sometimes means handing over funds to *de facto* government agencies (Hearn 2004b, 210, 220). Cuban authorities were tolerant of foreign intervention because, first, it was kept quiet to avoid awkward questions about the intersection of sovereignty and dependency; and second, it was accommodated into a Gramscian interpretation of "mixed" civil society amenable to state hegemony (Acanda 1997). Like food sovereignty, civil society's rhetorical currency as a guarantor of participatory development established consensus between Cuba and its donors. With its precise meaning never pinned down, civil society's unchallenged "frozen state" distracted attention from diverging interpretations of the term among community, government, and foreign actors (Hernández 2003, 122).

Reflecting the regulated autonomy of Cuban NGOs, the Foundation for Nature and Mankind (subsequently renamed the Antonio Núñez Jiménez Foundation, FANJ) became a prominent partner for international cooperation. It embraced the permaculture movement, which since its establishment in Australia by Bill Mollison and David Holmgren (1978) had developed practices and principles to achieve ecologically sustainable food production in cities around the world. Cuban families and community gardeners participated in FANJ permaculture training courses taught (and funded) by the Australian Agency for International Development (AusAID) and later sponsored by the United Nations Development Program, Oxfam, and others. Permaculture and agroecology were adapted to local conditions and refined, influencing the movement's global trajectory and giving rise to a powerful narrative of Cuban food sovereignty premised on ecological sustainability and political defiance. The AusAID program director in Havana noted the strategic use of this narrative, telling me in 2001: "They're running with the ball here, and although I'm not always comfortable that they're turning community gardens into political projects, in the end it's up to them how they want to go forward" (for the full case study see Hearn 2008, 111–122).

To "promote the sale of food products," the government's 2011 *Lineamientos* economic guidelines (sections 178–84 and 309), alongside Decree Laws 259 in 2008 and 300 in 2012, authorized private investment in commercial horticulture and permitted the acquisition of usufruct (freely leased) land between 13.4 and 67.1 hectares (República de Cuba 2011, 26–27, 38). By 2010, approximately 383,000 urban farms had been created on 50,000

Figure 4.2
Fresh food for sale at a Chinese-sponsored *organopónico* in the Havana suburb of El Vedado, 2018. Photo by the author.

hectares of land, supplying 70 percent of fresh fruits and vegetables consumed in Havana and other cities (Altieri and Funes Monzote 2012, 23–24). The lived implications of these reforms show themselves in the peri-urban farms that produce Havana's fresh food, where personal loyalties and values coexist with the state's pursuit of market socialism. Those who work these territories attempt to reconcile longstanding agrarian livelihoods with the changing demands of a city in the throes of reform. For Elena's brother Miguel, food was more than a product: "My family has used organic methods to feed Havana since the time of my great-grandfather. . . . We know this land better than the bureaucrats" (interview, June 20, 2018).

Miguel's statement supports a point argued by anthropologists Josephine and Alan Smart: Food is "emotionally loaded" with cultural identities and territorial affinities; compared to other commodities, "food can evoke everything from disgust to nostalgia for traditional lifestyles, emotions that are more visceral, deeply held, and easily mobilized" (2016, 99, 103). Cuban discussions of food sovereignty reflect this observation through the term's mobilization for different reasons in the paddock, at the market, and around the conference table. For instance, some farmers associate food sovereignty with independence not only from foreign suppliers but also from the Cuban government, which has struggled to facilitate access to machinery and other inputs (Leitgeb et al. 2016, 420–421; Thiemann and Spoor 2019, 546). Miguel seemed to share this view; when I asked him about the meaning of food sovereignty, he asserted his individual rights: "Well, that means I have the right to create and sell my products however I choose."

Contrasting with earlier quantity-based approaches, the 2020 Cuban Plan for Food Sovereignty and Nutritional Education (SAEN) "aims to encourage a culture of food and nutritional education for a healthy population" (Díaz Pérez et al. 2021, 10). Similarly, the National Assembly's 2021 draft Food and Nutrition Sovereignty and Security Law defines food sovereignty as "the nation's capacity to produce food sustainably and give the whole population access to sufficient, diverse, balanced, nutritional, safe and healthy food, reducing dependency on foreign means and inputs, while respecting cultural diversity and environmental responsibility" (República de Cuba 2021, 3). The emerging commitment to ecological and cultural diversity faces a dilemma well known to food sovereignty researchers: "It is difficult to draw fixed lines to separate what is 'culturally appropriate' and might be permissible within a food sovereignty paradigm and what is not. . . . Food cultures are a bigger barrier to food sovereignty advocacy than is often appreciated" (Edelman et al. 2014, 916–917).

One barrier to the Cuban government's more culturally sensitive and health-oriented framing of food sovereignty arises from Miki and others who prefer soft drinks to fruit juices and "don't eat vegetables." A wealth of ethnographic analysis shows that these and related choices are entrenched in long-running preferences that are fortified by memories of scarcity during the Special Period, when subsistence horticulture became a matter of survival. Mona Rosendahl observed that in the midst of crisis, providing *comida fuerte* (strong food)—such as rice, beans, and some form of meat—became a valued aspect of women's identities as wives, lovers, and mothers (1997, 36, 49–50). In the early 2000s, Anna Pertierra witnessed the legacies of this association: "Memories and experiences of the previous decade in which even the most dedicated, inventive, and determined cook may not have been able to produce good enough food, or even just enough food, is a trauma that threatens one's purpose in life" (2012, 50). The construction of *comida fuerte* to counter scarcity with alterity—and shame with pride—harks back to early Cuban expressions of independence from Spain, when white *criollo* farmers distinguished themselves from bourgeois city dwellers by embracing rice, beans, and pork as a marker of national identity (Moon et al. 2022, 74; Padrón Hernández 2012, 166).

To capture the power of memory in Cuban food preferences, Hannah Garth focuses on adequacy rather than sufficiency because, "the framework of adequacy can account for what is necessary beyond basic nutrition, prompting us to ask not whether a food system sustains life, but whether it sustains a particular kind of living" (2020, 7). The expansion of Cuba's private sector has provided avenues for Cubans to redefine adequacy and distance themselves from memories of scarcity, including through the consumption of previously unavailable foods. Miki's cheeseburger is a good example: Liberated from fresh vegetables produced on government subsistence farms and served in styrofoam used only by private restaurants, it appears to be a new form of *comida fuerte*. His diet suggests a nexus of past constraint and present possibility: "The eating associated with previous periods of scarcity," writes Garth, "may be characterized as a form of binge eating" that is satisfied by proliferating fast food options (2013, 90). Also grounded in history and evident in Miki's diet is elevated sugar consumption, reflecting a preference noted by food writer Fernando Fornet Piña (2015, 8) for soft drinks over fresh fruit juices despite the wide availability of the latter since the Special Period. If SAEN and the new Food and Nutrition Sovereignty and Security Law are to steer citizens away from such choices, they will need to turn the cultural tide on negative perceptions of fresh food.

The emerging food sovereignty policies align with the boom of "agroecotourism" to community farms and *organopónicos* since the Obama administration reestablished bilateral diplomatic ties and eased US economic restrictions. Obama's reforms, writes Richard Feinberg, opened the way for US citizens to travel and "carry any amount of cash for donations to Cubans . . . to spur the development of private businesses" (2016, 153). From the Cuban government's perspective, any influx of travelers and resources is welcome if harnessed to support state authority but prone to risk if diverted into the island's expansive informal sector. The calculation of this risk accommodates the less insidious nature of Cuban informality compared to other contexts, such as *blat* in the former Soviet Union, which gave rise to wealth-seeking criminal networks (Ledeneva 1998). It also differs from Chinese *guanxi* networks, whose family and political underpinnings constitute less accessible circles of reciprocity (Smart and Hsu 2004; Hearn 2016, 14–16). Cuban informality, by contrast, functions largely to meet (*resolver*) the daily needs of the general population and has been permitted to expand—particularly in the orbit of tourism—in ways that align with state interests (Bastian 2019, 186). "Socialist entrepreneurs," for instance, connect foreign visitors (and well-resourced Cubans) with medical services and supplies that "actively contribute to the daily functioning of Cuba's socialist health care system" and ultimately generate revenue for the state (Brotherton 2008, 260, 268).

From health clinics to hotels, the informal networks underpinning Cuban efforts to *resolver* daily needs and *luchar* (struggle) through scarcity intersect with broader political structures. As Kathy Powell writes of Cuban exchange networks, "The 'street's' informal practices are thus inherently, and intimately, political, and while they by no means exhaust the field of political relations, they constitute an important element of the state's ensemble of relations" (2008, 184). Powell also notes informality's political toll as "loyalty to the state [is] patchily eroded" (2008, 192) through what Archibald Ritter and Ted Henken call a "culture of illegality" (2015, 203). Food demonstrates this erosion, for instance, through the common practice of government bakers selling flour among networks of informal bread makers and consequently producing small loaves (Bastian 2019, 112–113). Chiseling through a grainy bread roll from the neighborhood state bakery in 2017, the father of my Havana host family shared his dismay: *"De pinga este pan, igual que el país"* (literally: This bread comes from the dick, just like the country).

The balance of power between formal state institutions and informal initiatives requires the exchange of information and resources between the two, giving rise to individuals who serve as go-betweens. Cuban intermediaries, aptly named "amphibians" by Margaret Crahan and Ariel Armony (2007, 2), are integral to the state's "ensemble of relations" as Powell suggests, but they share the professional precarity of those introduced in previous chapters. Among the consequences of their job insecurity are self-serving exploits, such as charging informal commissions from hotels, clinics, and restaurants, earning them the unflattering title *jineteros*: jockeys who ride their horses (i.e., tourists) for personal gain. Also like their foreign counterparts, Cuban intermediaries attempt to build bridges between formal and informal networks whose inward-oriented bonds do not naturally favor cooperation.

The following case studies suggest that those who forge such connections first research the agendas of prospective partners and then reconcile these through historicized narratives that resonate across divides.

Agroecotourism generates welcoming habitats for amphibians, furnished with opportunities to steer foreign revenue into the public, private, and informal sectors while boosting collective enthusiasm about Cuba's fresh food policies. Resulting outward-facing accounts of Cuban food sovereignty dwell on the lessons generated by the Special Period about *organopónicos*, (peri-)urban agriculture, and fresh food, attracting foreign visitors interested in these themes. Recent research by Krystal Moon et al. (2021, 32, 2022, 74) indicates that this interest is rubbing off on Cuban host families and communities, which are beginning to adopt healthier diets as their clients seek out salads and fruit juices. Such encounters advance the government's emerging food sovereignty policies, push for tourism, and vision of fledgling private businesses injecting resources into underserved communities. As the first case study shows, with benefits like these on the table, local officials may justify resulting informal exchanges as the inevitable efforts of their communities to *resolver*.

While agroecotourism aligns with the nutritional and environmental focus of recent food sovereignty policies, Cuban researchers have long advocated greater responsiveness to local interests and opportunities. An enduring agenda is to identify where excessive top-down control has impeded food producers' agency and "replace it with an approach that grants more authority to the productive bases" (Labrador Machín et al. 2020, 12). Writing in 2006, Cuban legal scholar Avelino Fernández identified a pressing need for "policies [to] align with the natural activities of food production, of people in the countryside, and their associative forms" (2006, 143). Such an alignment would give farmers sufficient autonomy to access stable supply chains and to determine their own balance of production and sales while ensuring customers safe, nutritious, and affordable food. "Only then," writes Abel Gamboa Costa et al., "would it be possible to achieve the benefits of a participatory principle of food sovereignty" (2013, 13). The push for greater self-sufficiency and autonomy among food producers, supported by farmers like Miguel earlier, has arguably deepened since the COVID pandemic. Agronomist Luis Vázquez (2020, 3) suggests that COVID hastened the need to refocus food sovereignty away from state-led import substitution and toward community-led agroecology:

> The reinforcement of food production in response to the COVID-19 crisis, if undertaken with a systemic focus rather than as import substitution, can help to realize transformations necessary to achieve socioecological resilience in Cuba and other countries facing climate change. Agroecology promotes food sovereignty because as well as guaranteeing the population's nutritional security it means augmenting capacity for self-supply in the territories with minimal external supplies.

Such arguments add contemporary nuance to earlier formulations of food sovereignty developed to break Cuba's dependence on the United States and to confront the shortages of the Special Period. In October 2021, Cuba's José Martí International Journalism Institute acknowledged the need to shift away from top-down approaches in a series of "tips for journalists" writing about food sovereignty. The advice includes "avoiding triumphalist language," "taking into account dependency on foreign imports," and "including diverse voices" (IIPJM 2021). These suggestions contrast with earlier official assertions of food sovereignty as a triumph of national self-sufficiency with little need for foreign support.

For their part, international NGOs and educational travel services, no longer sequestered from contact with ordinary Cubans as they were in the early 2000s (see an account in Hearn 2008, 113–114), have perceived the island's approach to food sovereignty as a model charged with insights for other countries and communities (FAO 2014, 10–19; Wright and Morris 2015). "The world," writes the Cuba-US Agroecology Network, "has much to learn from Cuba's journey toward agroecology and food sovereignty" (CUSAN 2020). The US nonprofit organization Cuba Platform (2020) observes that this journey has intensified during COVID-19:

> Before the coronavirus crisis began Cuban president Diaz-Canel called for food sovereignty and the Cuban Ministry of Agriculture announced new plans to increase production and limit imports. Now, with COVID-19, the need has become even more urgent. . . . There are a multitude of farmers in every city and rural area across the island who have sustained the country during challenging times; and who can show us all a road map for getting through this crisis.

Shortly before the pandemic, foreign programs invited visitors to "acquire the knowledge and strategies to create just, sustainable, local, and healthy food systems in your own communities" (Altruvistas 2019), offering Cuba as "an example to the rest of the world in sustainable agriculture" (CERES 2019, also see Global Exchange 2019). Solidarity associations arranged visits to Cuban horticulture facilities, where participants could "experience agricultural life" (ACFS 2019) and "see Cuban sustainable farming practices first hand" (CAFS 2020). In the pandemic's wake, these services are reigniting their collaborations with institutions like FANJ, whose Sustainable Local Development coordinator Caridad Cruz traces foreign enthusiasm to the resilience of the Special Period:

> Our food policies were strategic for Cuba in the 1990s, a time when we were practically abandoned by the world, and they are still important today because they are reconnecting us with the world. Look at the success of our "Toward a Culture of Nature" conference [noted earlier] and the number of foreign groups that come each year. It's really very clear: Cuba's experience is really relevant to everyone who believes in food sovereignty. (Interview, June 28, 2018)

Cruz observes that Cuban food sovereignty, as presented in academic conferences, solidarity programs, and agroecotourism, appeals to foreign environmental and social sensibilities. From this perspective, Cuba's experiences and innovations simultaneously fulfill two provisos for effective food system intermediaries: policy frameworks that support land, labor, and environmental reform, and historically accrued background knowledge that resonates across a wide "symbolic ecology" (Descola 1996). The following case studies suggest that the alignment of Cuban food sovereignty narratives with the progressive visions of what Gürcan (2014, 130) calls "international political society" is not coincidental. On the contrary, in both cases intermediaries function as *bricoleurs*, priming prospective visitors to find common cause with urban farmers, neighbors, and local officials through the benefits of nonindustrial fresh food. The rhetorical glue that holds together resulting networks is a concept of food sovereignty that accommodates each of these groups' interests and agendas. Semantic flexibility—common among US community farming activists according to Davenport and Mishtal (2019)—enables them to historicize food sovereignty in ways that strengthen bonds within these distinct groups as a prelude to building bridges between them.

Intermediaries at Work: Two Cases

My introduction to two illuminating scenarios resulted from my involvement with the Antonio Núñez Jiménez Foundation (FANJ) since 2001, initially as a PhD researcher and later as a coordinator of projects funded by the Open Society Foundations and the Australian Department of Foreign Affairs and Trade. My familiarity with FANJ staff and their activities favored efficient data collection and diminished the need to establish trust anew, but also increased the risk of insensitivity to evolving opportunities and constraints (c.f. Adler and Adler 1987, 47–48; Couser 2004; Hayano 1979, 99–101; Shim 2018, 4–5). Erroneous assumptions, getting caught by surprise, and other displays of naiveté were more common than I like to admit, but these awkward experiences (some of which appear later) helped to fill gaps in my understanding of how Cuba is changing.

I had come to know the two project leaders, Luis and Reinaldo, as loyal employees of FANJ and an associated urban farming organization, and while they remained so, they were now taking advantage of economic reforms to pursue more independent interests. Reflecting Adriana Premat's (2012) observation of a growing distance between state ideologies of collective endeavor and individual entrepreneurship, they now fit Crahan and Armony's description of "amphibians" (2007, 2). The implications of their deepening independence dawned on me when a casual coffee meeting with a FANJ program director in 2019 turned into an interrogation about their activities. The director's questions suggested discomfort that Luis and Reinaldo may be transgressing their professional limits by working with foreign visitors, local governments, and entrepreneurial neighbors. As network theory predicts, their "strong ties" to FANJ aroused the possessive pull of "negative social capital" (Portes 1998, 15).

As I drafted the case studies the COVID-19 pandemic struck, intensifying Cuba's economic and nutritional challenges. In the first two months of 2020 tourism had declined by 16.5 percent, and by year's end was halted by closed borders (González 2020). Moreover, COVID's impact in Florida and elsewhere cut remittances, a vital source of income for many Cuban families. Prior to the pandemic these amounted to around USD $6 billion annually, including goods acquired during trips abroad and (in 2018) USD $3.7 billion sent in cash (Morales 2019). Taking advantage of newly introduced smartphone internet connectivity, entrepreneurs in Havana offered a workaround for the government's "stay home" mandate by using WhatsApp to advertise home delivery of fruits, vegetables, meat, and fully cooked meals for those who could afford them (Bastian and Garth 2020).

A text message from a friend in Havana described COVID's impact on the ground as "depressing . . . nobody is working and some of our neighbors don't have anything to eat" (October 27, 2020). Luis and Reinaldo remained active during this time, expanding their community welfare activities while reaching out online to foreign contacts for support. As Reinaldo put it in an email, "The memory of the Special Period is present like a ghost, though it seems things won't get so bad this time. Biden's victory is giving us hope in a resumption of the friendly policies of Obama's final years" (November 25, 2020). By 2024, hope had evaporated that the Biden administration might roll back Donald Trump's policies, such as designating Cuba as a "state sponsor of terrorism." Associated restrictions on travel, remittances, and foreign collaboration remained in effect, contributing to the island's inability to meet production targets for staples including eggs, milk, and meat (Suárez Fernández 2024). Deepening nutritional deficiencies provoked the Cuban government to request children's powdered milk from the United Nations World Food Program. During five weeks in 2024, I witnessed the scarcity of basic foods and the inability of Luis and Reinaldo to relaunch their projects. Reinaldo offered a sober analysis:

> It's now almost impossible to import animal feed, and young Cubans are leaving to the US because they get favorable residency treatment there, which has shrunk the workforce. So, we're not able to produce food, and on top of this, the inflation that came with the unification of our currency has put the price of basic things like a pizza at over 2,000 pesos in private restaurants. With the current restrictions it's too complicated for me to organize visits from US universities. (Interview, September 17, 2024)

Collaborating with Australian institutions proved simpler: Between 2022 and 2024, Reinaldo facilitated a project supporting medicinal plant production by the Afro-Cuban organization Ifá Iranlowo, funded by a grant I secured from the Australian embassy's "direct aid program." Exercising characteristic skill, he orchestrated a signing ceremony where I could discuss the project with FANJ's senior staff, lawyers, and technicians. I invited Ifá Iranlowo's leaders to this and other meetings, whose encompassing purpose was to build a participatory process for working out goals and ensuring collective satisfaction that inputs—farming equipment, transport, and the time of all concerned—were adequately paid. The project's progress is being recorded on my website (SOM 2023) and will likely feature in future publications. Its development suggests, like the below case studies, that Reinaldo is conscious of the potential conflicts of interest raised by the FANJ program director, and that he has no intention of distancing himself from his official employer. On the contrary, he and Luis place high value on the opportunities that arise from their jobs, such as opportunities for foreign engagement.

In the case studies to which we now turn, Luis and Reinaldo employ strategies to consolidate their occupational status. Professional stability is a critical yet often elusive proviso for intermediaries, whose methods include emphasizing their projects' benefits to their employers. An inclusive approach to cooperation therefore permeates their work: supportive of institutional aims, aligned with government policies, and animated by personal goals. This positive-sum vision enables them to strengthen bonds within distinct interest groups as a basis for building intergroup bridges, in this way transcending the constraints of negative social capital.

Case Study: The Eco-Lodge

Amid the worn colonial architecture of Central Havana, the Eco-Lodge is attracting international attention. For three decades its director, Luis, has introduced foreign NGOs and universities to the horticultural innovations of Cuba's urban farms, and in return they have hosted his presentations around the world. Having converted his deceased parents' home into a licensed three-bedroom hostel with around 1,200 square feet of vegetable plots crammed into the courtyard and rooftop, he now offers his audiences an opportunity to visit Cuba and put his lessons into action. Central Havana is a largely Afro-Cuban suburb that reflects the reemergence of racial disadvantages since the 1990s. As Katrin Hansing and Bert Hoffman (2020) find in a survey of more than a thousand people across Cuba, the gap is growing between people of African and European ancestry in access to remittances, salaries, savings, internet access, foreign travel, and high-income private activities such as property rental. Luis says that to build urban farming projects that improve local conditions requires international connections:

> Generally, everyone is in favor of sustainability and community development, but they don't always agree about how to achieve them. So, I explain to everyone that, during the Special Period, the blockade [US embargo] obligated us to protect our food sovereignty . . . Everyone knows that we the Cubans suffered a lot, and that creating new techniques permitted us to survive. The techniques functioned, and still function, to benefit of our communities, but we need resources from outside to maintain them. (Interview, March 18, 2018)

Luis shares his message with foreign travelers, who gather around the Lodge's permaculture display on a cool autumn evening in 2019. Nine visitors pass around a basket of assorted lemons, tomatoes, and onions as Luis invites them to "smell the lemons and take a tomato if you like; these are the benefits of the Special Period that nobody talks about." On the front porch of a gray concrete house across the street, retired engineer–turned-horticulturalist Federico sways vigorously in a rocking chair and shares his observations:

> Last week on *Ecos* [a television program] they mentioned that people around the world praise our food sovereignty in Cuba because it got us through the shortages by producing food without chemicals. I'm not surprised that the tourists come to learn about this, and I hope more of them will stay at my house and see my garden too. I don't want to depend on them or anyone else, but this is a good opportunity for us and Luis is helping as a friend. . . . You know, in Cuba our businesses have been restricted for a long time, but more and more the *Lineamientos* [economic reforms] have made it easier to compete *como quiera* [however one wants] to increase our income. (Interview, September 28, 2019)

Hoping to tap into the Eco-Lodge's appeal, Federico implies that shifting economic policies are unleashing new opportunities and competitive pressures, not least through the *Lineamientos'* promise to connect emerging small businesses with tourism lodging and gastronomy (República de Cuba 2011, 33–34). Such intersections have become common as agroecotourism initiatives offering "high-touch" experiences stimulate satellite services ranging from meals and coffee to boarding and cooking lessons (Moon et al. 2021, 22). For Federico the advent of this niche is opportune: His sister and two nephews moved away in 2017, leaving two bedrooms available for renting. One of these is covered by a *casa particular* license whereas the other is unregistered, and both have been in high demand from Luis's visitors. In return for "helping as a friend," Luis charges Federico a commission as per normal in the informal networks that *resolver* daily shortages. This reciprocity, says Luis, orients his larger mission:

> If we are creative then we don't have to rely on the government for our food sovereignty. . . . The project is helping the neighborhood to develop and demonstrates that it is possible to feed ourselves in a sustainable way. People are really interested in this because the whole community benefits, you see . . . little by little I'm creating a circuit of benefits with neighbors who can provide accommodation and get involved. (Interview, September 29, 2019)

Here Luis's invocation of food sovereignty implies greater local autonomy from the state, but it must nevertheless articulate with the requirements of the Popular Power (*Poder Popular*) political system. Like any city block (*manzana*), Luis's is overseen by a Committee for the Defense of the Revolution

(CDR). CDRs were once subsidized to ensure citizens' revolutionary loyalty and coordinate social welfare activities like school enrollment programs and blood drives (Eckstein 1994, 22). Post-Soviet austerity saw their social services decline, but they continue to monitor the legality of neighborhood affairs and provide character assessments of individuals seeking job promotions, defending court cases, developing small businesses, and so on. Luis is mindful of his CDR's scrutiny and often discusses the changing policy environment with its captain, Marta. He notes his recent debate with her about the relationship between *Lineamientos* 3 and 12. The former asserts that "in the non-state system the concentration of property will not be permitted," while the latter insists on "the absolute observance of legality" (República de Cuba 2011, 11). The ambiguity of "concentration" contrasts with the rigidity of "absolute observance," leaving Luis and others wondering to what extent they can expand their businesses without getting into trouble. A chat with Marta suggests that she and Luis have navigated the murky boundaries of formal business and informal exchange by focusing on shared values:

> We're aware of Luis's service to Cuba and we know he is developing this project for educational reasons. He has informed many people in other countries about Cuba's accomplishments through the difficult times, so they know that our country is a leader in food sovereignty. You've seen this in the way he teaches people about permaculture and sustainability, and also he brings income to the community. This is why we support him. . . . Even the North Americans find ways to come here to learn and what do they see? They see that we are well, in spite of the blockade and their President Trump. More and more of them keep coming to see projects like this, so one can conclude that more and more people know the truth. (Interview, September 30, 2019)

Marta has led the CDR since retiring from her job as a clerk with the local *Consejo Popular* (municipal government council) six years prior. Her professional transition, provoked by her husband's death and her daughter's increasing work responsibilities, has enabled her to dedicate time to caring for two grandchildren and for her neighborhood. She says Luis has always been well known and liked in the community, as one of the few who "comes and goes" through international work that advances the revolution's mission. Aware that her endorsement alleviates many of the uncertainties he faces as a private operator with an expanding project, she identifies common cause in the Eco-Lodge's commitment to community education and economic uplift, as well as its defiance of US pressure.

Recognized as an outstanding CDR captain by the *Consejo Popular*, with the associated certificates and photographs displayed on her worn living room wall, Marta has achieved results despite the scarcity of centrally allocated resources. Like her counterparts in other CDRs, she is encouraged to support projects that generate law-abiding economic opportunities for her territory (c.f. Hearn 2008, 97, 115–117). The Eco-Lodge serves this purpose, enabling her to spread the word about Cuba's food sovereignty achievements while advancing the government's push for neighborhood initiatives that produce income from foreign tourism rather than consuming government subsidies. With these benefits at stake, Marta overlooks the informal character of Luis's network, demonstrating official tolerance for the need to *resolver* daily economic challenges in areas that align with—rather than erode—state authority.

While Marta and Federico see political and economic merit in the Eco-Lodge, foreign visitors focus on its ecological contribution. The sixteen travelers I speak with either heard about the project through international environmental groups that previously hosted Luis's presentations or personally met him on those occasions. Arriving at the Lodge independently of institutional exchange programs or study tours, they contribute informal donations and pay Luis in cash to arrange classes and accommodation with his neighbors.

As sunlight filters into the leafy courtyard on a summer morning in 2018, the visitors stand up to stretch and chat after the morning activity, scheduled on the noticeboard as a "presentation on the history of permaculture." Leaving aside the wanton observation that the class only covered 1989–1995, I solicit reflections on their motivations for visiting. Arkansas student Rebecca states that, "I don't want it to be all about money, but I'm happy to help out with the work they are doing, and I've come here to learn from them about how to live sustainably. If I can be useful by contributing money, it's the least I can do." In the same courtyard a year later, Arandra, a nurse from Melbourne, tells me she communicated frequently with Luis for six months to learn about the Eco-Lodge and arrange her stay. When I lament that my emails have often failed to reach his inbox, she exposes my ignorance of recent mobile internet reforms: "Why don't you join his WhatsApp group?" Thankfully she moves the conversation on: "We [the WhatsApp group] share knowledge about Cuba, because there's a lot of misinformation spread by people who are against Cuba. . . . What they're doing here goes way beyond food production because they're creating a community that's sustainable."

Rebecca and Arandra's social and ecological sensibilities are distinct from, yet compatible with, Federico's entrepreneurial interests and Marta's patriotism. Each perceives tangible expression of their values at the Eco-Lodge, whose pursuit of food sovereignty affirms internal bonds among visitors, neighbors, and local officials. Adapted by Luis to align with specific agendas, the term's meaning is broad enough to furnish a bridge between them. Yet the portrayal of food sovereignty as a fixed tradition tied to the Special Period cuts history short. Emerging challenges, such as intensifying commercial competition brought by recent economic reforms, the resulting reintroduction of agrochemicals, and tight government controls on imported equipment, get little attention. In this regard Luis differs from a second intermediary, Reinaldo, who explicitly links the island's historical achievements to current structural problems. The theme of food sovereignty is similarly prominent in his work, but it is articulated both to honor the past and critique the present. Inward bonds consolidate around this more critical and unfolding vision of food sovereignty to generate corresponding outward bridges, producing a network whose diversity, stability, and formality are quite different.

Case Study: The Funes Family Farm

In the rolling hills of Caimito, between Havana and Mariel, the Finca Marta (Marta's Farm) is pioneering new opportunities in Cuba's emerging mixed economy. Its agroecological methods produce organic fruits, vegetables, and honey for local schools and community centers, private restaurants, and clinics specializing in health tourism. The enterprise is managed by researcher-turned-farmer Fernando Funes with support from his father Fernando Sr., an eminent agronomist, and his brother Reinaldo, an environmental historian working at FANJ.

Reinaldo has deepened the farm's international connections through his visiting professorship at Yale University, where he teaches and runs study tours to Cuba. In 2018, at FANJ headquarters amid Amazonian pottery collected by the Foundation's intrepid originator, he tells me about the tours:

> Critical thinkers in Cuba and the United States want to connect with each other, and many misunderstandings can be cleared up when people see reality on the ground. Yale is progressive in this way and that's why we visit the *Finca*. . . . If we simply rely on the market, then imports from the United States could unfairly compete with our local farmers. Our economy is opening and I think our government

> should take action to regulate artificial pesticides and fertilizers. . . . Cubans can learn from our own experiences of food sovereignty and environmental management, and I've proposed setting up a Special Period Research Center to do this. The tour shares these ideas. (Interview, June 29, 2018)

The daughter of the Foundation's legendary creator and its current director, Liliana Núñez, joins us and affirms Reinaldo's vision: "Food sovereignty means avoiding dependence, and Reinaldo's examples show very clearly that to achieve this we must rethink our policies. Part of this is financial, so in our projects, like in Sancti Spíritus province, we're pushing for greater economic decentralization." The disposition to recognize emerging problems alongside past achievements is apparent in an academic conference on Cuba's historical ties with China organized by Reinaldo at Yale in 2019, followed by a study tour to Havana. Presentations on Cuba's deepening economic integration with China and past reliance on sugar exports to Spain, the Soviet Union, and the United States become the backdrop for a firsthand look at food sovereignty in action at the *Finca*.

As I am in the United States en route to Cuba, I eagerly accept Reinaldo's invitation to join the conference and tour. Awaiting our delayed flight to Havana with coffee on the table and time to kill, one of my fellow travelers, a historian, offers an opinion on Cuba's challenges:

> Their [Cuba's] traditions are about fighting for independence. It was interesting what Reinaldo said, that they're trying to protect their sovereignty from the negatives of globalization, so they have to grow their own food and work with allies like China. . . . If you stop to think about it, if China is anything to go by, you know, the way they pump chemicals into their food, then they [Cubans] might want to think twice. (Interview, September 26, 2019)

The historian perceives that Cuba's pathway to food sovereignty may be undermined by the government's choice of allies and resulting damage to the food system. Critiques of counterproductive policies reemerge four days later, as the group takes cover from the tropical sun in the shade of the *Finca*'s patio. Gesturing to a bowl of fresh burgundy tomatoes on the table, Reinaldo's brother Fernando explains that the meaning of food sovereignty has evolved:

> You can't produce vegetables that taste and smell like this with chemicals. This farm could rapidly increase production with chemicals like some of our competitors, but we don't agree with this shift because Cuba has a comparative advantage that we want to protect. . . . There is pressure from everywhere to increase profits, but I think our customers recognize our natural methods as a strength and part of

> our food sovereignty. The state sometimes doesn't see this. . . . During the Special Period the state helped us to create new organic techniques, and farms provided job security and direct access to food. But now you see the situation has changed, and now young people prefer to live in the city, so we guarantee workers a salary that increases every year, and we employee fifteen people from the area. Hiring them has been complicated and we want reforms in that area. . . . The new constitution [ratified in 2019] should help us because for the first time since the 1960s it recognizes the role of the market economy in national development. But there's a lot of room for improvement to protect organic farmers, like letting us import tractors, greenhouses, and water tanks. (Personal communication, September 30, 2019)

Like Miguel (Elena's brother), Fernando associates food sovereignty with greater economic and territorial autonomy, but he is uncertain about the future: On the one hand his vision is undermined by economic competition, while on the other, it may benefit from more liberal trade and employment policies. Juan Machado, a farmer retained when Fernando acquired usufruct rights to the property from its previous operator in 2011, says the brothers have established good relations with neighbors by helping them to face emerging pressures. I approach him and four of his team as they observe us from benches next to a vegetable field. When our conversation turns to food sovereignty he gestures, gloves in fist, to his companions:

> Look, you probably know that we went through hard times in the Special Period, and the most important thing we can offer is the lessons we learned. The community here didn't know the Funes family before and was suspicious [*tenía dudas*], but I told them we have to adapt to stay employed. . . . So, we do the planting, digging, and building, and the [Funes] family is showing people [neighbors] how to do it on a big scale organically and with methane energy and new technology. (Interview, September 30, 2019)

For Machado, food sovereignty entails new exchanges to manage challenges as they arise. But when the tour group meets Reinaldo's colleagues at FANJ headquarters, questions arise about the management of exchange. I record the encounter in my research diary:

Cuban Consensus

The FANJ international coordinator welcomes the group to the conference room, whose walls are an archive of Antonio Núñez Jiménez's published and unpublished manuscripts, maps, and diaries. Her Yale counterpart reciprocates: "We're delighted to work with you. What are your connections like with other universities?" Reinaldo interjects: "We are a bridge into Cuba so we receive students from everywhere, even from China." Two of the visitors suggest Taiwan and Hong Kong as more environmentally responsible partners than China, and they reiterate this several times during the meeting. Like the FANJ officials, I remain silent about the importance of Chinese trade and investment for Cuba and the conditions this brings. Reinaldo changes the subject, proposing a possible line of cooperation: As an NGO, FANJ is authorized to receive US students and equipment for its education programs. The proposition is welcomed as a way to collaborate into the future while circumventing legal restrictions on both sides.

Research diary, Havana, October 31, 2019

Figure 4.3
The Finca Marta, 2019. Photo by the author.

The excerpt shows Reinaldo steering the discussion from discord to consensus, shifting away from politically sensitive opinions about China's growing influence to shared interest in Cuba–US cooperation. While food sovereignty is not the explicit focus, it permeates years of preparation for such encounters. At FANJ Reinaldo has emphasized the concept's relevance for reforming Cuban environmental and economic policies and for building propitious external partnerships, goals that resonate with his colleagues' professional mission. At Yale he has articulated food sovereignty as a tool for critical reflection on Cuba's development, a proposition that advances commitments to innovative teaching, research, and engagement with the island. His interpretation of the term thus strengthens bonds within each group and, against the political odds, underpins bridges between them. The endeavor is vitalized by the *Finca*'s neighbors, who Machado describes as proponents of a pragmatic exchange. Trusting his endorsement, they bring experience and muscle to an enterprise that, in return, sustains their rural livelihoods. Inward bonds stimulated to inspire outward bridges again generate a corresponding network. The chapter's final section finds causality in this chain of meaning: the telling of tradition performed by intermediaries directly influences the membership, stability, and formality of the networks they build.

Time and Ties

To build projects, intermediaries need networks, and to build networks, they need history. To this end food sovereignty is a valuable rhetorical resource, sufficiently global to resonate across different social groups while local enough to accrue specific meanings for each. As Don Nonini writes, "Food may have multiple meanings at one time, or shift in meanings over time, through the nature of the exchanges in which it is involved" (2014, 399). Examples abound: Farmers in Belize have adopted collard greens, lettuce, and other salad ingredients into their "food culture" to feed "hungry gringos" (Griffith 2018); family farmers in central Cuba have begun to eat mixed salads after foreign tourists requested them (Moon et al. 2021, 27); and in Papua New Guinea's coffee industry, "the livelihoods of importers and roasters depend on having the commodity carry a different symbolic meaning from the one attributed to it by people living and working [there]" (West 2012, 13). Such intersections of local food systems with international markets suggest that even when chains of production are stable and fixed, "chains of meaning" evolve (Osterhoudt et al. 2020).

Embodied in the tomatoes, onions, and educational offerings of Havana's urban and peri-urban farms, Cuban food sovereignty exhibits multiple chains of meaning, whose evolutions have given rise to contradictions. As the opening anecdotes show, the antilogies include growing uptake of chemical pesticides and fertilizers as lamented by Elena, deepening reliance on foreign support despite the assertions of state officials, and unhealthy food choices that diverge from official (and foreign) advocacy of local fresh food production. That many Cubans reject fruits and vegetables in favor of *comida fuerte*, which now includes cheeseburgers and the novelties of an expanding private fast-food industry, helps to explain why the government's emerging food sovereignty policies emphasize nutritional health.

The recent policies reflect Cuba's changing economy, in which Chinese support and agroecotourism have yielded new revenue but deepened old tensions. Rebuilding the export sugar industry alongside domestic rice and corn processing strengthen national capacities but depend on foreign finance. The underlying paradox of food sovereignty reliant on external patrons surfaces at the Eco-Lodge and the Finca Marta, which demonstrate the concept's ambivalent relationship with international webs of trade and tourism. But resulting uncertainties about the concept's meaning do not undermine its power; rather, they provide fertile ground for intermediaries like Luis and Reinaldo to imbue it with fresh emotive associations. As Barthes observed, broadly familiar signifiers—whether an image on the cover of *Paris Match* or a caption at the entrance of Elena's local market—are constantly refilled with new meaning. Luis and Reinaldo provide this meaning, harnessing the wide appeal of food sovereignty in their own ways to build common ground between communities that may otherwise have little in common.

The purposeful articulation of Cuban food sovereignty invites anthropological inquiry into a conceptual puzzle that runs through this book: how do intermediaries reconcile the tension between internal bonding and outward bridging to build networks? Unravelling this puzzle requires insights into *why* networks form to begin with, drawing attention to the proactive construction of consensus. Luis and Reinaldo furnish such insights, evoking the locally poignant yet globally familiar theme of food sovereignty to foment bonding prior to bridging, and moreover, historicizing the concept as a Cuban tradition to shape their networks.

Luis built the Eco-Lodge to embody food sovereignty as a tradition forged from the pains and innovations of the Special Period. By associating the project with an emblematic historical moment, he attracted a wide variety of visitors, who traveled independently to Cuba for a range of professional, academic, and personal reasons. The opportunity to personally experience the fruits of Cuba's historic food crisis appealed to Rebecca, Arandra, and others, who had learned of the Eco-Lodge through first- or second-hand connections with Luis. Neighbors like Federico catered to foreigners interested in Cuba's post-Soviet survival story by providing classes and accommodation, though the payments they received implicated them in an expanding informal network. CDR captain Marta perceived no threat to state authority in this effort to *resolver* daily economic needs, endorsing the Eco-Lodge to showcase her neighborhood's resilience "through the difficult times." These formative interactions constituted what Luis called a "circuit of benefits" around the Eco-Lodge that was contingent but not coincidental. The diverse range of network participants, the short time frame of their association, and the informality of their transactions bespoke a carefully historicized telling of tradition.

Reinaldo's historical vision of food sovereignty contrasted with that of Luis, drawing together professionals interested in both the lessons of the Special Period and the problems of the present. Acknowledging the accrual of systemic failures through time, he emphasized Cuba's colonial and post-colonial dependence on sugar production for export to Spain, the United States, and the Soviet Union; the government's outdated employment and trade regulations; and the resurgence of agrochemicals amid competition brought by economic reforms. This dynamic approach to history appealed to foreign scholars researching Cuba's international relations and long-term development, establishing a point of convergence with FANJ's progressive mission. In background and purpose, constituents of the FANJ–Yale partnership were more homogenous than Luis's varied visitors, and better supported to sustain collaboration into the future. Institutional formality manifested in the structure of reciprocity: A detailed itinerary of activities was contracted and paid prior to departure, ensuring compliance with US and Cuban legal requirements for study tours. The give-and-take played out on the ground among neighbors, who explained their "planting, digging, and building" as an adaptation to waning job prospects and emerging opportunities. Food sovereignty, envisioned as a critically reflexive and unfolding tradition, gave rise to a network marked by restricted membership, long-term stability, and formal exchange.

The formation of networks around the Eco-Lodge and the Finca Marta contribute nuance to the literature on intermediaries outlined in the introduction. While ethnographers have shown that intermediaries draw on broadly familiar values and traditions to diminish uncertainties and establish "working consensus" (Goffman 1969), Luis and Reinaldo reveal the consequences for network formation. By historicizing an esteemed Cuban tradition to align with the sensibilities of their interlocutors, they galvanized inward bonds to underpin outward bridges, generating networks that supported their goals. Their endeavors suggest that intermediaries seeking to achieve quick results, particularly under rigid political conditions, are likely to invoke a static tradition from a patriotic moment in history. Conversely, if they seek long-term systemic change, then an evolving tradition premised on historical continuity is more fit for purpose.

Although packaged in the same rhetorical signifier (food sovereignty), the distinct meanings signified by fixed and flexible temporalities appealed to different groups, whose collaboration in the case studies generated corresponding network structures. The former network was diverse, diffuse, and informal, enabling community action while flying under the political radar. The latter was closed, stable, and formal, permitting reflexive critique while fomenting institutional collaboration on emerging food system challenges. By stimulating such interplays of cultural affinities and network structures, Cuban intermediaries continue to inflect and influence food sovereignty's meaning at home and abroad.

Amid growing international concerns about industrial agriculture, environmental impact, and dependency on China, food sovereignty's importance is widely agreed. That it is invoked in response to such a range of concerns demonstrates both its global relevance and local resonance. Cuban articulations of the term substantiate Philip McMichael's point that each of its valences is "governed by a different order of historicization" (2014, 952). As noted, it has been advocated at distinct historical junctures to reject colonial and postcolonial dependency, invoked in urban agriculture programs to inspire post-Soviet resilience, marshalled in policies to promote nutritional health, and advertised in emerging agroecotourism projects to celebrate—and sometimes critique—Cuba's track record. Food sovereignty's diverse deployments have, over time, generated domestic contradictions whose persistence continues to inspire international interest.

The unresolved tensions of Cuban food sovereignty manifest in political and cultural uncertainties. Among those noted in this chapter are the degree of economic autonomy available to private food producers like Elena and Fernando; the blurred boundaries between private, state, and informal actors; and the paradox of national food sovereignty reliant on foreign support. Chinese cooperation with Cuba has provoked concerns about new dependencies, not least in the export sugar economy, but both sides have steered the bilateral relationship away from Soviet-style patronage. Cuba has instead engaged the global tourism industry, within which agroecotourism aligns with policy priorities but exposes a rift between foreign perceptions of Cuban food sovereignty and reality on the ground. Community-grown fresh produce does not figure prominently in daily diets, and Miki's love of soft drinks and cheeseburgers—conditional on the extraction of offending vegetables—reflects the growing availability and appeal of fast food. Ethnographic accounts suggest that the strength of this appeal is fortified by memories of scarcity during the Special Period and by longer-running habits including disdain for uncooked vegetables and embrace of sugar. The ingredients of *comida fuerte*—rooted in national independence and intensified after the Soviet collapse—appear to be evolving once more as the expanding private sector serves up enticing but unhealthy choices.

Food sovereignty's uncertain meaning has given rise to intermediaries, whose articulations of cultural values restore a measure of shared purpose across divides. Digging into the term's rich cultural undersoil, Luis and Reinaldo brought its accrued emotive power to bear on practical pursuits. Broad familiarity with food sovereignty, seared into collective memory through the harrowing quest for nutrition under siege, was a foundational proviso for *bricolage*: the creation of projects that marshalled everyday discourse about a known Cuban tradition to address emerging needs. It was no coincidence that one intermediary explained the formation of this tradition as a fixed moment and the other as a continuing process, as distinct temporalities oriented their projects and networks. For Luis, the guiding agenda was an immediate "circuit of benefits" for his community, while for Reinaldo the goal was long-term progress toward sustainable farming and associated reforms.

Although their agendas and historical orientations diverged, both intermediaries consolidated their job security and efficacy by aligning their projects with the broad priorities of Cuba's command economy. Among these priorities are the development of a small and controlled private sector to improve conditions in underserved communities, broader education about nutritional health, and promotion of agroecotourism. These principles fulfilled another proviso for Luis and Reinaldo: a supportive policy environment that enabled their diffusion of narratives about Cuba's food sovereignty achievements and potentials, generating corresponding networks at home and abroad.

By invoking the cultural past to shape networks, Luis and Reinaldo illuminate an interdisciplinary concern. Network sociology provides a conceptual framework of bonding and bridging but does not explain how inward affinities are transformed into outward consensus, nor how the motivations that underpin this process influence network characteristics. Ethnography elucidates the centrality of intermediaries to such conjunctures as they reinforce values within groups as a prelude to orchestrating collective pursuits. Just as a concert master requests the orchestra's woodwinds, strings, and brass to tune to "A" before harmonizing with each other, intermediaries tune their interlocutors to a common cause prior to collaboration. This chapter has suggested that food sovereignty is one such cause, historicized by intermediaries to strengthen intragroup bonds prior to building intergroup bridges. The "multiple sovereignties" that arise from such mediations provoke a practical and theoretical question: "Who is the sovereign in food sovereignty?" (Edelman et al. 2014, 919–920). This chapter has suggested that *who* the sovereign is depends on *why* they join networks, and that intermediaries are instrumental in amplifying and harmonizing their motivations for doing so.

Unfolding challenges vitalize food sovereignty with urgency and appeal. In Cuba, the concept's circulation by officials, farmers, tour operators, and intermediaries underscores a common message: The food system innovations that saw the island through the post-Soviet crisis are now more relevant than ever. As food supplies fluctuate in the wake of COVID, a question posed by policy analyst Ming Chun Tang (2016) is resurfacing: "Will a pioneering, sustainable food system that emerged from a period of extreme scarcity and hardship survive the transition to an era of relative abundance?" Formulations of food sovereignty, tailored to blend personal agendas with structural conditions and collective sensibilities, will guide the answer to this cyclical question. Attempts to achieve quick results within prevailing political structures will deepen trends already in play, potentially including greater agrochemical use to increase harvests. Alternative strategies that advocate long-term solutions will critique prevailing structures to push for economic and ecological reforms. Intermediaries will cultivate both scenarios as they sow networks and narratives from the seeds of an evolving tradition.

Taking stock of the tension between short-term gains and long-term reforms, the final chapter considers the interplay of growing overseas agricultural investment, particularly from China, with community concerns about territorial sovereignty, environmental stewardship, and nutritional health. The intermediaries introduced so far demonstrate that strategically articulated food traditions can provide common ground for building networks that voice and advocate these concerns. How, then, might the rising superpower cultivate its own intermediaries to reach out and build consensus about the future of food systems?

5 Bridging the Belt and Road

Inside Knowledge

A red-and-gold name card, rather elaborate for a one-on-one meeting, indicates where I should sit. Across from me, also behind a name card, my host apologizes for arriving late: A television news team in the adjoining room had unexpectedly requested an interview on the Belt and Road Initiative. The back wall of the meeting room glows soft azure, as if to emphasize the importance of the thirty or so scarlet folders neatly arranged along it on glass shelves. "Welcome" says my host, a professor of public administration at Renmin University of China. "You are the first foreigner since COVID to visit our new headquarters."

Inwardly lamenting the less-adorned meeting rooms I'm accustomed to, I recall that Renmin University is different even among its national peers. The first university to be founded by the Chinese Communist Party (CCP), it bears the cultural esteem of "first son," whose senior officials are said to have the "red gene" [红色基因, *hongsi jiyin*].

When our conversation turns to the role of Chinese researchers, especially those with the "red gene," in advising their government on policy matters, he gestures to the red folders neatly lining the illuminated wall. "These are our *nei can* [内参, literally "internal reference"] reports. They contain policy suggestions, and after months of review and revision, they go to the Communist Party Central Office. After more revision and redaction, some are sent up to the Central Committee led by Xi Jinping. My recent *nei can* was used in the Central Committee's new policy on urban sustainability. . . . I know this because I recognized some of the text. It's a detailed process and not for open publication—even the author's name is removed." He lifts the ceramic lid from a tall teacup adorned with traditional nation-paintings [国画, *guó huà*] and takes a long sip. Another nice adornment, I muse, and do the same.

(continued)

Inside Knowledge (continued)

After our meeting he accompanies me out of the building, and as we say goodbye, I congratulate him for his *nei can*. He breathes deeply and looks to the sky. "Thank you," he sighs with a long exhale, "but you have no idea . . . I hope it's the last one."

Beijing, September 15, 2023

Perplexed by my host's ambivalence, during September and October 2023, I asked policy researchers and advisors in Beijing and Shanghai about the mechanisms they use to conduct and communicate their work. This chapter draws together resulting observations with others gathered over two decades, during a cumulative eighteen months in the two cities. It aims to clarify the causes and consequences of recent tensions in China's relations with its suppliers of agriculture commodities, foregrounding the personal perspectives of Chinese researchers tasked with advising their government through mechanisms like the *nei can*.

The preceding chapters have observed that China's demand for natural resources correlates with its urbanization agenda, evident at the Wang family's Beijing dinner table (chapter 1) and in the central government's policy to increase the national urban population to one billion people (Xinhua 2014). Underpinned by an ambition to "facilitate economic development by boosting domestic consumption," the policy is on track to convert 350 million rural farmers into urban consumers by 2030 (Chu 2020, 1).

China's chief planning bureau, the National Development and Reform Commission, has repeatedly stated over the past decade that food security and equitable urban growth are among the nation's most pressing challenges for the twenty-first century (NDRC 2013, 2023). Its 2023 annual report outlines a "holistic approach to national security" in which food security is consistently listed in first place as a target for the government's total annual investment of 680 billion RMB (USD $93 billion; NDRC 2023, 22, 44, 53). This commitment aligns with public expectation: In a survey I conducted with Cuban sociologist Rafael Hernández of our fifty-six students at Renmin University, thirty-four strongly agreed and seventeen agreed with the statement that "It is the government's responsibility to make sure people have access to healthy food" (Hearn and Hernández 2018).

From ancient times until the Second World War, China produced more than 85 percent of the world's soybeans. Since then, technological innovation in the United States and Latin America, and China's 2001 entry into the World Trade Organization, have led to ever-increasing imports of this fundamental source of protein. Moreover, the crop's intensive use of water and farmland, both of which are scarce in China, has prompted farmers to shift into more profitable corn production (Xinhua 2013). Government investments in agricultural technology and mechanization have increased self-sufficiency in corn, rice, and wheat production, but in the soybean sector, these cannot not overcome the forces of global trade and domestic geography. Han Han, a scholar of Latin America at the Chinese Academy of Social Sciences (CASS), suggests a parallel with Mexico:

> Soybeans are native to China, and 3,000 years ago, we were the first to convert soy from a wild plant into a domesticated food. This is why soy milk, oil, and tofu are part of Chinese traditional culture. It's sad to see how we must now import soy, especially since most of it is genetically modified. It is a sadness we share with Mexico and its traditional corn. The increasing trade data may look beautiful, but the consequences for our farmers and our cultures are far from it. (Interview, October 18, 2023)

To secure China's food supply, state-owned enterprises (SOEs) in the agriculture sector have "gone out" to invest overseas like their counterparts in the energy and mining industries before them. The activities of SOEs such as COFCO, Heilongjian Beidahuang Nongken Group, Xinjiang Production and Construction Corps (XPCC), and Chongqing Grain Corp aim to provide food at stable prices, which have proven no less volatile than for other commodities. Producing and procuring foreign soybeans advances this cause, but it also requires Chinese enterprises to learn about unfamiliar environments and interact with societies that are not always receptive.

A confluence of pressures stands in the way of Chinese ambitions to expand agriculture and other extractive operations overseas. Outlined later, these include progressive community-led objections to social and environmental risks, which have killed proposals to open new soy production and associated infrastructure projects in Brazil. They also include trepidations about industrial competition and economic dependence, analyzed by researchers since the early 2000s when Chinese demand for Latin America's resources ramped up (Dussel Peters 2019; Hearn and Manríquez 2011; Myers and Wise 2017; Strauss and Armony 2012). Incompatible priorities are a prominent theme in this literature, especially in the agriculture sector: While supplier countries want to upgrade their agribusiness industries by attracting foreign capital into infrastructure, food processing, packaging, and other higher-value-adding segments of the food chain, Chinese investors have prioritized the production and extraction of raw commodities. Contributors to these books point out that a lack of Chinese interest in purchasing soymeal, soy oil, and other products derived from crushed and processed soybeans has entrenched a classic pattern of dependency on highly mechanized primary exports, which fails to add value and generate employment.

Public debates in the Americas, Australia, and Africa register concerns about the local impacts and economic outcomes of Chinese projects but also about their political implications. Investing firms stand accused of acting in the service of the Chinese state and, therefore, avoiding the rules that regulate market competition. In the words of Sergio Amaral, former chairman of the China-Brazil Business Council (CBBC), "Sometimes you don't know whether the investments are looking for Brazil as a market or whether they correspond to strategic purposes of the Chinese government." Chinese outbound investment surged during the 2003–2013 presidency of Hu Jintao, provoking the CBBC (2011, 26) to articulate a global regulatory challenge:

> This issue is controversial in Brazil and other countries (Canada and Australia, particularly) and has led governments to take action. In this respect, it seems that a consensus is being formed: countries need strong legislation and institutions which can clearly distinguish and characterize opportunities and threats arising from the sale of land to foreign groups.

The "strong legislation" advised by the CBBC has manifested in caps on foreign land purchases in Argentina, Brazil, and Australia. In some cases, criticism has been openly hostile, invoking the specter of "neocolonialism" to associate Chinese investment with a "mercantilist approach" reminiscent of prior foreign exploitation (Camus et al. 2013, vii).

How Chinese planners learn about and respond to social, economic, and political concerns in supplier countries is poorly understood. This chapter aims to shed light on this process, starting with an account of the deep tensions that complicate Chinese projects in its most important soybean supplier, Brazil. It then considers China's Belt and Road Initiative (BRI) as an emerging, but as yet unrealized, response to these tensions and comparable concerns in Australia and Cuba. While geopolitical and macroeconomic arguments loom large in most accounts of BRI, a culturally oriented approach more fully accommodates the initiative's origins, aims, and capacities to engage the world. I suggest that BRI has set out to first build broader domestic familiarity with China's heritage as a pioneer of international integration along the ancient Silk Road, and to then project this narrative outward.

As with other repurposed traditions, described in previous chapters, intermediaries are central to BRI's historicization as an extension of Chinese cultural heritage. The final section considers the ethical complexity faced by Chinese intermediaries and foreign researchers, like me, who critically engage with their attempts to reimagine the past to shape the future. Their work is conditioned by the same provisos navigated by the book's other intermediaries—enabling government policies, public familiarity with tradition, and supportive professional environments—but as the chapter's opening anecdote suggests, mediating between state and society in China poses unique challenges. The chapter concludes that the way Chinese intermediaries balance inward bonds with outward bridges, particularly in relation to the ethics of professional status, is shaping BRI's projection and reception.

Decolonizing Public Opinion in Brazil

When I first met Jiang Shixue in 2007, he was a researcher in the CASS Institute of Latin American Studies, focused on China's bourgeoning trade and investment relations with the region. During the intervening seventeen years, his publications, presentations, and interviews have conveyed creative and frank reflections on the associated political and social dynamics of these relations. Now director of Shanghai University's Center for Latin American Studies, he maintains this approach:

> Be careful what you believe about the Belt and Road Initiative, because people like to exaggerate. Recently, when Brazil's President Lula visited Beijing, a Chinese scholar wrote that BRI would significantly increase bilateral trade. But in reality, although BRI might create long-term stability, it would mainly be a political gesture because there is nothing stopping trade. It's huge already, and actually, some people think it's too big. (Interview, October 17, 2023)

Jiang's point aligns with Brazilian research on BRI, which finds that joining the Initiative would be "a symbolic act" for Brazil rather than an economic one (Ferreira Abrão and Amineh 2024, 1). And yet Chinese trade with Brazil, underpinned by the exchange of the former's manufactured goods for the latter's natural resources, has been accompanied by Chinese proposals to purchase, lease, or otherwise assume control of Brazilian farmland. Anxiety about this prospect is often expressed through websites, social media, and offhand remarks like that of former Brazilian minister of finance Antônio Delfim Netto, quoted in chapter 1: "The Chinese have bought Africa and now they're trying to buy Brazil." Reported in mainstream media, Delfim's comment attracted broad support at a time of rapidly growing Chinese investment (e.g., Estadão 2010), prompting the federal government to formally revive a national land registry system and introduce a 5,000 hectare (and maximum 25 percent) limit on foreign land investments. Similarly in Argentina, opponents of a bid by Chinese firm Heilongjian Beidahuang Nongken Group Co. to rent 320,000 hectares to grow soybeans in Río Negro province used social media to describe the project as a "neo-colonial landgrab" and successfully pressured the Argentine congress to cancel the contract. Foreign land ownership across Argentina was subsequently limited to 1,000 hectares, not to exceed 15 percent of any property.

Rodrigo Maciel and Dani Nedal (2011, 250) suggest that Brazil's legislation goes beyond the protection of national interests:

> Sinophobia has also played a part in recent legislation limiting land purchases by foreign companies and individuals. Chinese FDI is said to be qualitatively different from that of traditional sources because of the controlled and opaque nature of the Chinese economy, China's selectivity in allowing inbound FDI, and the close association between investing companies and the Chinese state.

Negative perceptions of China contrast with the earlier enthusiasm of Brazilian politicians and industry leaders toward Chinese finance during Hu Jintao's monumental 2004 Latin American tour, when international media reported his promise to invest USD $100 billion in the region by 2010. Between 2005 and 2013, Brazil received USD $31.4 billion of Chinese investment, making it the fourth largest destination for Chinese finance after the United States, Australia, and Canada, but two decades after Hu's visit, Chinese investment remains overwhelmingly focused on energy, metals, and agriculture. Because it depends on raw soybean exports, the expansion of Brazil's agriculture sector does not signify progression up the value chain. As a policy paper from Kings College and the Brazilian Center for International Relations concludes, "Brazil is China's main source of agricultural imports, with an import market share of over 20 per cent. Despite these impressive numbers, Brazil has not been able to satisfactorily implement its declared official priorities in relations with China: diversification and increasing the share of high value-added products in Brazil's exports" (Rosito and de Carvalho 2022, 4).

Brazil's colonization by Portugal and later relations with the United States entrenched a disadvantageous pattern of cheap commodity exports in return for expensive manufactured imports. Conscious that China may be cast in the same colonial light, analysts in both countries who favor closer Sino-Brazilian integration emphasize BRI's anticolonial agenda. For instance, in a recent collection of papers on Brazilian perceptions of BRI published by the Getúlio Vargas Foundation, Rodrígues and Jabbour (2023, 135) write that the initiative is "an expression of a modern geopolitics that is popular, anticolonial, and nationally liberating, whose treatment of diverse civilizations substitutes distance for exchange, confrontation for mutual learning, and feelings of superiority for peaceful coexistence." Anticolonial narratives are also common among Chinese commentators, who cite their nation's experiences of European colonization as a basis for empathy with Brazil and Latin America. Jiang Shixue argues that China's relations with the region should be understood as South–South cooperation because, unlike colonial predecessors, China has inflated rather than depressed commodity prices and reduced rather than increased the price of manufactured products and capital goods. Furthermore, he writes, "while colonial powers sought to monopolize markets by discouraging the development of industries in their colonies, China invests actively in technology transfer programs, which have assisted the development of local industries across the continent" (Jiang 2011, 62–63).

Greater transfer of value-adding technology to Brazil's agriculture sector would, according to Rosito and de Carvalho, build "a higher degree of bilateral trust that could contribute to long-term stability in food exports" (2022, 16). Successive bilateral agreements have sought to facilitate such transfers, but most have been frustrated by cultural and political misunderstandings. A good example is Chongqing Grain Corporation's 2010 agreement with the government of Bahia state to build a soybean processing complex near the town of Barreiras. The proposed USD $300 million investment was to be the first of six facilities (totaling USD $2 billion) that would crush soybeans to produce soymeal and soy oil, adding value in Brazil rather than in China. The plant would employ local workers, source soybeans from the immediate region around Barreiras, and reserve a proportion of the soy oil and meal it produced for the local market. By promising local benefit and added value, the project distinguished itself from previous colonial endeavors and attempted to alleviate concerns about the negative implications of Chinese investment.

As an incentive for the investment, the mayor of Barreiras provided Chongqing with 100 hectares, adding to the firm's bid for 100,000 hectares of farmland (CBBC 2011, 26; Powell 2011). But the powerful landless rural workers' movement (MST) criticized the scale of the acquisition, joining a loose coalition of civil and environmental advocates to describe it as a land grab. Chongqing Grain Corporation's status as a Chinese government enterprise provoked suspicion that it may not act according to market principles. As the newspaper *Estadão de São Paulo* put it at the time, Chinese state-owned enterprises "may act according to commercial interests, like other investors, but may follow the logic of a state—and not the Brazilian state" (Estadão 2010).

Brazilian anxiety, accompanied by federal limits on foreign land ownership, have bemused Chinese officials such as Zhang Dongxiang, CEO of the Bank of China's Brazil branch: "Public opinion sometimes seems to be against foreign investment . . . as if it makes local industry less competitive. . . . These are some antiquated ideas" (quoted in Winter and Stauffer 2013). The Barreiras project was shelved, having received approximately 15 percent of the projected USD $2 billion by late 2014, and most of this in assessment and mapping activities.

As the Barreiras project lost traction, Chinese President Xi Jinping visited Brazil in 2014 with a proposal to invest USD $8.6 billion, including the Trans-Oceanic Railway project to transport soybeans and iron ore across Brazil and Bolivia to Peru's Pacific coast. Promising to introduce logistics technology developed by the China Railway Eryuan Engineering Group, the project has not materialized because, according to Brazilian researchers, the Chinese side "failed to grasp the political, environmental, and economic complexities involved" (de Sousa et al. 2023, 503). Others blame exaggerated Brazilian expectations about Chinese capacities to coordinate ecological and political approvals across borders (Schutte 2020), as well as incompatible Chinese contracting rules that allow providers to be appointed with no competitive bidding process (Rosito 2020). Jiang Shixue has a different explanation:

> Chinese companies are pretty good at listening to communities, but it is up to Latin American governments to engage in a productive way with their own societies. That's not China's job. So, when the community groups protested about the railway's possible environmental and social problems, what did the Chinese

> companies do? They said, "Well, if you don't want it, then we'll go away." And they did. (Interview, October 13, 2023)

Attempts to build goodwill through technology transfer remain prominent in Chinese accords with Brazil, including those signed during President Luiz Inacio "Lula" da Silva's 2023 visit to Beijing. According to Brazil's official list of "signed acts," these include investment in value-adding processing of commodities and in "sustainable rural production, especially for agricultural families and communities" (Governo do Brasil 2023). Despite the potential benefits of these and other projects, they cannot avoid being associated with the interests of a foreign government.

The Chinese government's proactive role in foreign trade and investment is defended by Professor Wang Yiwei, vice president of a Renmin University research institute whose name recalls the "red gene" discussed earlier: the Academy of Xi Jinping Thought on Socialism with Chinese Characteristics for a New Era. Wang argues that Chinese SOEs have the long-term vision and capacity needed to address the world's infrastructure gap, and that these strengths should generate a welcoming environment in Latin America and elsewhere (interview, September 11, 2023). One factor that has impeded this welcome, argues Professor Zha Daojiong of Beijing University, is a misinterpretation of China's SOEs:

> I think there's a connection drawn in many countries between the way Chinese SOEs work and how Western state-backed corporations behaved in colonial times. However, our SOEs are not the Dutch East India Company! For one thing, ours try to localize and learn about the contexts we're in, and they also observe how Western companies operate in China and follow their examples. (Interview, September 27, 2012)

While Chinese SOEs may differ from their Western antecedents, it is unlikely they will be embraced in Latin America or anywhere else until they demonstrate tangible benefits on the ground. The Confucian precept of trust as a precondition for working together may not extend as widely as expected, for as David Shambaugh (2009) observes, "for the Chinese, cooperation derives from trust—whereas Americans tend to build trust through cooperation."

To accelerate the formation of trust, Chinese SOEs are pursuing a second strategy: Rather than build it from scratch, buy existing stocks. To this end, they are relinquishing their pursuit of land acquisitions and focusing instead on mergers and acquisitions that yield controlling stakes in established agribusiness networks. Early signs of this strategy emerged in 2014, when China's largest agriculture SOE, the China National Cereals, Oils and Foodstuffs Corporation (COFCO) spent USD $1.2 billion to acquire 51 percent of Dutch conglomerate Nidera, which has major agriculture holdings in Argentina, Brazil, and Eastern Europe. Four months later, COFCO committed USD $1.5 billion for 51 percent of the agribusiness unit of Noble Group, a Singapore-listed commodities trader that operates grain warehouses and loading stations in Argentina, Brazil, Uruguay, and Paraguay and five processing plants in Asia. In 2021, as the COVID pandemic deflated asset prices, Chinese investment in Brazil compared to 2020 more than doubled to USD $5.9 billion to reach a cumulative USD $71.6 billion. New projects by Chinese carmakers BYD and Great Wall Motor are reportedly taking over production plants vacated by Ford and Mercedes (Myers et al. 2023).

Buying into locally trusted enterprises and picking up where other foreign firms left off reduces the specter of external threat. For COFCO this advances an ambition to compete with the world's "Big Four" multinational grain brokers (Archer Daniels Midland, Bunge, Cargill, and Louis Dreyfus). Leaving executive boards, administrative structures, and employment practices largely unchanged further diminishes perceptions of economic aggression.

Investing in infrastructure and acquiring existing agribusiness networks may diminish criticism of Chinese enterprises, but concerns about the growing influence of their government persist. Fueling these concerns is a broader narrative, embraced in liberal democracies but officially rejected in China, about the dangers of government intervention and the merits of private markets. "Legal apparatus," writes Fukuyama, is a "substitute for trust" because state interference in economic affairs "crowds out" the natural inclination of actors to cooperate with each other (1995, 27). This argument helped to consolidate a US-led vision of globalization after the Cold War, which Fukuyama famously celebrated as "the end of history" (1992). However, China's emergence as a superpower, underpinned by assertive policies that prioritize the stability and preeminence of the state, shows that Fukuyama's declaration was premature. History is still being written, and the Belt and Road Initiative is putting government intervention back on the agenda for partners around the world.

Balancing the state with the market has been a longstanding point of contention among Brazilian food security advisors. Many lament the continuing legacies of colonization by Portugal and later relations with the United States, which entrenched a disadvantageous pattern of cheap commodity exports—most recently soybeans—in return for expensive manufactured imports. The need to escape from this value-eroding predicament was the thrust of Raúl Prebisch's (1950) seminal book *The Economic Development of Latin America and Its Principal Problems*, which encouraged Brazil and its neighbors to pursue import-substitution industrialization from the 1950s until the 1980s. The continuing influence of dependency theory on governance in Brazil has underpinned longstanding academic and executive consensus that the state should be entrusted to manage key aspects of economic development (e.g., Cardoso 2001; Cardoso and Faletto 1979; Oliveira 2006).

The *Lava Jato* (car wash) scandal that dominated Brazilian politics between 2014–2020 demonstrates that the betrayal of public trust has deep consequences. An initially minor investigation into money laundering at a car wash in Brasília evolved into a landmark anticorruption probe of the nation's largest companies and state-owned enterprises. Millions of ordinary citizens congregated in the streets to protest nepotism and inadequate transparency, leading to the 2016 impeachment of President Dilma Rousseff. Her replacement by conservative lawyer Michel Temer, and the subsequent election of Jair Bolsonaro, only deepened public disillusion—nationally and around the world—as the agribusiness sector expanded into the Amazon. In 2017, Temer lifted a moratorium on deforestation to permit private cattle ranching and soy cultivation in an Amazonian area the size of Denmark, and in 2022, Bolsonaro abolished local-level indigenous committees and openly advocated land clearing to further expand farming in the region (Cardoso Carrero et al. 2022; Gro Intelligence 2017).

Lula's 2023 reelection as Brazil's president represents a return to more progressive social and environmental policies, but his capacity and disposition to reign in agriculture developers in the Amazon, the Cerrado Savanna, and elsewhere is far from certain. I asked Evandro Carvalho, director of the Center for China-Brazil Studies at the Getúlio Vargas Foundation, if Lula can stand up to Brazil's powerful agriculture sector. His response emphasized the link between unrestrained soybean exports to China and disinterest among agribusiness leaders in adding value to these exports:

> Agriculture is more powerful than any politician in Brazil, even the president. Most of the players in our soy sector are from the United States and Europe, and they're aligned with the right wing of Brazilian politics. So, there's an alliance of business leaders that is quite happy for Brazil to continue exporting raw soybeans. The way they see it, if we were to refine large quantities of soybeans into soy oil and soy

> meal, it could diminish their business and profitability. It's actually very similar to the oil sector, where we export massive quantities of crude oil but do very little refining to add value, and this has generated all kinds of corruption, like in the *Lava Jato* scandal. (Interview, October 19, 2023)

Carvalho and I met in Beijing during the Third Belt and Road Forum, while participating in a television talk show on the global impact of China's "modernization" (CGTN 2023). The show's host had asked us to reflect on the legacies of European and US hegemony in developing countries, and how the Belt and Road Initiative (BRI) might usher in a "new path of modernization for all." Referring frequently to foreign colonialism, his questions established a historical context upon which BRI could improve. Among my responses to these questions—edited out of the broadcast but included in my research diary—was that modernization was itself a colonial ideology, and that Chinese overseas projects might benefit from postmodern recognition of ethnic plurality, gender equality, civil society, and diversity in general. After the recording the host reminded me that although BRI embodies the spirit of China's ancient Silk Road, it is young and in the process of consensually formulating best practices for international engagement. This, he said, is why foreign commentators were invited to contribute our suggestions in the televised discussion. Before considering some of the ethical dilemmas arising from participation in such activities, the next section explores the role of Chinese intermediaries in articulating BRI's origins and aspirations for global audiences.

Building the Belt and Road

The Belt and Road Initiative (BRI), known in China as One Belt One Road (一带一路, *yi dai yi lu*), is President Xi's signature platform for augmenting outbound investment in infrastructure and facilitating the exchange of Chinese manufactured goods for foreign agricultural, mineral, and other resources. During the decade since its inception in 2013, the initiative has attracted 152 countries, accounting for more than a third of global trade, two-thirds of the world's population, and USD $966 billion in construction and investment (Baker McKenzie 2017; Ehteshami et al. 2024; Hillman 2018, 1; Leahy et al. 2023). The proliferating English language literature on BRI from policy think tanks, news media, and academia has evaluated BRI's growing economic significance, geopolitical weight, and environmental impacts. Critics allege that the initiative advances Chinese geopolitical influence and empowers authoritarian regimes, while laying a "debt trap" for countries that cannot afford to pay back BRI loans (Chaudhuri 2018; Lloyd 2018; Losos et al. 2019). A World Bank working paper titled "China as an International Lender of Last Resort" describes a system of "Bailouts on the Belt and Road" that has enabled Argentina, Pakistan, and others to avoid defaulting on their debts to the International Monetary Fund (Horn et al. 2023, 4).

The Chinese government describes BRI as a mutually beneficial investment framework that prioritizes environmental responsibility and social impact. Examples are outlined in a white paper called *The Belt and Road Initiative: A Key Pillar of the Global Community of Shared Future*, published by the State Council in conjunction with the October 2023 Third Belt and Road Forum. The document summarizes BRI's achievements to date, from "adding vitality to economic globalization" through railways, power plants, and agriculture infrastructure to "improving the mechanisms for people-to-people exchanges" through theater performances, museum exhibitions, and Confucius Classrooms. This comprehensive range of global and local contributions, says the paper, represents "a new paradigm for state-to-state relations that shapes the international order towards greater justice and equality" (PRC State Council 2023, 20).

In his speech to the Third BRI Forum, Xi (2023) pledged a further 780 billion RMB (USD $107 billion) and signaled the initiative's increasingly "small yet smart" orientation: "China will carry out 1,000 small-scale livelihood assistance projects, and enhance vocational education cooperation through Luban Workshops and other initiatives." The reference to Luban, a revered inventor and carpenter born in 507 BC, underscores BRI's growing focus on "people-to-people exchanges . . . to enhance dialogue on civilizations with BRI partner countries."

The 2023 BRI white paper notes that community building through people-to-people ties has been spearheaded by the agriculture sector, in which more than 2,000 Chinese experts have traveled to over seventy countries to share farming technologies techniques (PRC State Council 2023, 18). Focusing on knowledge transfer and community engagement avoids the financial risks borne by large infrastructure projects and, according to Maria Carrai (2021), demonstrates an emerging approach to corporate social responsibility (CSR) among Chinese enterprises to compensate for lax governance standards in host countries. A frequently cited "small but beautiful" project is a Chinese technique for cultivating mushrooms called *jūncǎo* (菌草, literally "fungus grass"), which has been adopted in 106 countries across Asia, Africa, Latin America, and the South Pacific. Endorsed by the United Nations Department of Economic and Social Affairs, *jūncǎo* draws on a traditional Chinese method of growing mushrooms for human and animal consumption in chopped grass rather than wood logs, enabling commercial scale without deforestation (UN-DESA 2021).

BRI's shift toward local engagement reflects the turbulent post-COVID economic environment, in which rising inflation and interest rates are making it harder for partner countries to repay their loans. According to Yunnan Chen, a researcher at the London-based Overseas Development Institute, big ticket infrastructure projects remain important, but "there has clearly been a pull back as well as a greater risk aversion. . . . I don't think this means an end to the BRI as a narrative. But the rhetoric has shifted already. We are seeing more mentions of small and beautiful projects" (quoted in Leahy et al. 2023). Official portrayals of BRI offer clues to its trajectory, prompting researchers to analyze its rhetorical depiction among diplomatic, academic, and industry networks. One recent book sets out to "interrogate the Chinese government's representation of BRI" (Sintusingha et al. 2021), while the aptly titled *One Belt, One Road, One Story?* argues that the initiative's "narrativization" is critical to its success (Miskimmon et al. 2021). Timur Dadabaev (2018, 31, 37) suggests that BRI narratives are premised on flexible notions of the ancient Silk Road to accommodate the diverse cultures and contexts of partner countries:

> The notion of the Silk Road has changed from a static concept of a historical trade route into a product of social construction upon which various states have built their relations. . . . The Silk Road is not a foreign policy doctrine but rather a discursive strategy of engagement that largely exists in the realm of narration. This narration is also a matter of social construction that is subject to change depending on the international environment. . . . The varying meanings of the "Silk Road" framing leave space for complementarity.

As a discursive strategy, the Silk Road invokes a distant past that is globally familiar yet sufficiently ambiguous to accommodate political diversity and temporal change. Like the historical narratives deployed in previous chapters, its various interpretations "leave space for complementarity" as actors with distinct agendas seek opportunities for mutual support. Consensus on BRI's legendary origins provides a cultural substrate for the construction of roads and railways to carry Chinese products, scientific methods, and farming techniques outward into the world. Xi Jinping (2023) articulated this temporal nexus at the Third Belt and Road Forum by emphasizing BRI's continuity with ancient Chinese history:

> Trains speeding along rail tracks, automobiles running on roads, flights connecting different countries, cargo ships breaking waves, and e-commerce bringing so much convenience to people: they have all become symbols of international trade in the new era, just like camel caravans and the sailing ships were for the past age. . . . I once said that the pioneers of the ancient silk routes won their place in history not as conquerors with warships, guns, horses or swords. Rather, they are remembered as friendly emissaries leading camel caravans and sailing ships loaded with goods . . . What has been achieved in the past 10 years demonstrates that Belt and Road cooperation is on the right side of history.

Convincing Chinese and foreign audiences that BRI offers a model of globalization inspired by friendly camel caravans from a bygone era requires more than presidential speeches. As the previous chapters demonstrate, reimagining the past to shape the future requires the connective skills of intermediaries.

The Chinese state has long cultivated and harnessed the semi-independent status of universities and think tanks to inform public perceptions. As the former director of Fudan University's Center for Latin American Studies, Zhu Hongbo, told me fifteen years ago, "One thing that we scholars believe is that it's our job to be intermediaries to explain ideas between the Chinese government and the population" (interview, November 26, 2009). This approach is evident in the creation of Belt and Road research centers within Renmin University, the Chinese Academy of Social Sciences, Shanghai University, and hundreds of other institutions. Tasked with studying BRI's development, hosting international conferences on its reception overseas, and promoting its expansion, these centers provide a nexus between state and society. Their scholars are sponsored to conduct research in BRI countries and present strategy suggestions to Chinese enterprises, as well as policy recommendations to the CCP Central Office through the *nei can* process outlined earlier. They are invited by the media, online and in print, to explain the Belt and Road's relevance to China and the world, and by extension, the importance of Chinese traditions to contemporary globalization. Formulating historically embedded narratives that resonate across cultural and political contexts is a not a byproduct of this work but rather its core purpose, and those who do it are intermediaries not by circumstance but by profession.

The institutional and economic resources afforded BRI intermediaries constitute an enabling policy environment, satisfying a key proviso for their effectiveness. The magnitude of this support dwarfs the progressive government grants and market opportunities sought by food system intermediaries introduced in previous chapters, including in Beijing. Yet government policies and funds gain little traction unless they are animated by an appealing collective goal. Professor Wang Yiwei of Renmin University, whose tireless media appearances have earned him the informal accolade "Mr. BRI," defines this goal as "the common rejuvenation of ancient civilizations" (quoted in Guangming 2023). The time and travel that he and hundreds of others dedicate to publicizing this goal and its material fulfillment inside and outside China suggests that a second proviso for effective intermediaries—public familiarity with a unifying tradition—is a work in progress.

An October 2023 afternoon with Jiang Shixue at Shanghai University offered an unexpected glimpse into public familiarity with BRI. As we traversed the sprawling campus on rental bikes, he pointed out colorful banners announcing the Third Belt and Road Forum. Arriving at the staff canteen, he shared some thoughts: "Everybody who is dining in here could tell you a thing or two about BRI and the Silk Road because Chinese scholars love connecting things with cultural traditions, but let's see normal people." He approached the kitchen area and asked a waitress if she had heard of BRI, which she had not. He then posed the question to a waiter and called through the service window to ask the kitchen staff, and then another waitress. Nobody had heard of BRI, though the chef thought it might be a men's clothing store.

Jiang's creative demonstration, albeit anecdotal, suggests that BRI does not resonate widely among the Chinese public. This suggestion stands at odds with official claims that the initiative is embedded in traditional Chinese culture, and it helps explain why the state invests in intermediaries to raise awareness about it at home and abroad. If BRI is to be successfully projected into the world, then like Cuban food sovereignty in chapter 4, it first requires a domestic platform. Recalling Putnam's seldom-cited point that inward bonding and outward bridging are not a zero-sum pair, the book's case studies have shown that intermediaries intentionally strengthen internal solidarity as a basis for building external cooperation. Just as urban farming and environmental stewardship serve as unifying traditions in the Brazilian and Australian examples, Chinese intermediaries emphasize the ancient Silk Road to build BRI's public profile.

To strengthen familiarity with BRI as an extension of Chinese tradition, Professor Wang Yiwei (2023) has published a children's book titled *Lights Along the "Belt and Road"* ("一带一路"的灯光). It is his tenth book on BRI, and the first by any author on the topic written specifically for a young audience. An article about the book in the Chinese online magazine *Business Weekly* suggest that a "transformation of tradition" is enabling the BRI to mature like a child into adulthood: "The Belt and Road Initiative emphasizes 'becoming oneself' rather than 'becoming someone else' and innovative transformation of the country's traditional culture. . . . Children are the new force and builders of the Belt and Road Initiative" (Xu 2023). In the article, Wang explains the goals that motivated him to write the book: "Not only should we make a Belt and Road book that is easy to understand and colorful to attract children, but we should also make the spiritual connotation of the Belt and Road Initiative flow silently into their hearts." The book features vibrant and fanciful pictures, such as a Chinese bullet train speeding past the Sphynx and other international icons; mythological fables from participating civilizations; and a "Father and Son Dialogue" section to guide family discussions.

Like his children's book, Wang's television appearances, social media posts, and news interviews—including the following—emphasize the bond forged from collective memory:

> The Belt and Road has activated history and culture, and everyone is looking for a common historical culture and historical memory, and suddenly they are looking for where Zheng He went in ancient times, where Zhang Qian went, surpassing "you and me" to explore "ours." It is a people-to-people bond for all countries that identify with the spirit of the Silk Road, and that weave a network of connected partners. . . . You cannot break the old system without studying history and culture. (Quoted in Guangming 2023)

When I asked Wang how awareness of history might "break the old system," he spoke of Cuba:

> The old system means outdated ideas. For example, you work in Cuba, where the United States still clings to historical memories of the Cold War. Some politicians in Washington, DC, even see Cuba as a new Pearl Harbor that is now under threat from China. Those people think like that because they are from a young country with few historical comparisons to make. But we in China understand what it is like to be colonized by a foreign power, so we try to help Cuba and other countries to break their dependency. That has been Cuba's problem throughout history, so

> we are helping Cuba to diversify its economy. . . . It's a delicate balance between economic opening and dependency, and diversification is key for independence.

Chinese activities in Cuba, carefully framed by both sides as examples of BRI's support for national sovereignty, include investments in the energy and agriculture sectors (chapter 4). CASS scholar Han Han suggests that historical solidarity has made China more willing than other countries to undertake such projects:

> Independence and sovereignty are priorities for Cuba, and the Cuban government requires all foreign projects to support these goals. Cuba recognized China diplomatically in 1960 and joined the Belt and Road in 2018, and because of our special relationship China has been willing to accept these requirements more than others. (Interview, October 18, 2023)

Articulations of historical solidarity are a mainstay of Sino-Cuban diplomacy, featuring in the statements of both countries' leaders whenever they meet. Among the examples are President Hu's 2004 statement that the two nations "are fraternal brothers . . . passing the test of changing and adverse international circumstances"; Raúl Castro's 2014 declaration that bilateral relations are "at their best in history"; President Xi's 2018 assertion that "Chinese people will never forget that Cuba, led by Comrade Fidel Castro, was the first nation in the western hemisphere to forge diplomatic ties with China 58 years ago"; and President Miguel Díaz-Canel's 2018 praise for Xi's "theoretical and practical contributions to the construction of socialism" (quotes in Jabbour 2022; Lam 2004; Wilson 2014; Xinhua 2018). Portrayals of harmonious bilateral ties avoid reference to less sanguine episodes, like Fidel Castro calling Mao Zedong a "senile idiot" in 1967 as the Sino-Soviet split forced Cuba to distance itself from China (quoted in Garcia Triana et al. 2009, xxi). Cheng Yinghong (2007, 728) argues that selective memories of Cuba carry value for the Chinese state:

> Cuba has been perceived as the most unyielding and thus the most admirable anti-American hero who, as the Chinese government has introduced to its people, sets an example of the defiance of "international pressures" and of survival against all odds. In this way the image of Cuba has facilitated China's newly emerging nationalism. The hardliners, old Maoists and new leftists have looked upon Cuba as the example of the socialism purer than China's.

Ma Guihua, who was appointed in 2016 as chief Cuba correspondent for China's leading newspaper *Xinhua*, confirmed Cheng's point when we met in Havana: "Cuba's story inspires Chinese people to improve their own circumstances" (interview, December 4, 2016). Through this rhetorical process, writes Cheng, Cuba has come to function as "a mirror for China's own reflection" because sensitive topics such as inequality and political dissent are not easily broached within China unless projected onto a foreign "socialist other" (2012).

China's "special relationship" with Cuba, as Han Han points out, has brought about initiatives that companies from other countries are unlikely to pursue. This was evident during a 2016 tour of investment opportunities arranged by the Cuban government for Australia's minister of foreign trade, Andrew Robb, which featured a stop at the Mariel Port and its newly adjoining Special Economic Zone. Timed to dovetail with President Obama's visit to Havana, the tour was co-organized by the Australian government's trade and investment division to furnish early insights into potential projects before they were snapped up by foreign competitors. I eagerly accepted an invitation to represent the higher education sector, and during the ensuing site visits and receptions I observed the impressions of my fellow delegates. Among these was a mining industry executive, who worried that "history might repeat itself; I mean, look what they did to the American companies." Appraisals from the tourism and renewable energy representatives were similarly glum, citing government restrictions on foreign ownership, hiring and firing, and financial flows. Underpinning these trepidations was a perception of overbearing state regulations and doubts that Obama would relax the US embargo enough for Cuba to ease them. While similar complaints are quietly shared by some Chinese executives (see Hearn 2016, 211–212), they have not impeded political and economic cooperation through BRI and other mechanisms. By contrast, in the absence of "fraternal" solidarity, none of the visiting Australian firms have pursued ventures in Cuba.

Just as Cuba's integration with China echoes the currents of historical diplomacy, international responses to BRI are conditioned by the dynamics of political alliance. The 152 nations that have so far joined the initiative are developing countries, many of them postcolonial, with tenuous relations with the United States. It was, therefore, unusual that in 2018, Australia's wealthiest state, Victoria, broke with national policy to independently join BRI. The episode underscores Australia's peculiar standing, discussed in chapters 1 and 2, as an advanced economy deeply integrated with US security and political interests and yet reliant on the export of commodities to China. Victoria's BRI agreement was criticized in the media and among conservative think tanks as a "zombie project" that would "allow the employment of an unlimited number of Chinese workers" and "make Australia part of a more China-centred world" (Lloyd 2018; Shoebridge 2020). Arguing that the agreement was "inconsistent with Australia's foreign policy," in 2021, Foreign Minister Marise Payne (2021) nullified it. Her decision conveyed a well-established narrative about Australia's distaste for "authoritarian models" and commitment to an "international order built around rules" (Payne 2020). This affirmation of the US-led liberal democratic order contrasts with the "people-to-people bond for all countries that identify with the spirit of the Silk Road" articulated by Wang Yiwei and other BRI intermediaries.

How BRI might integrate with—and improve upon—prevailing global norms was the theme of the television talk show mentioned above, whose full title was *BRI: A New Path to Modernization for All* (CGTN 2023). I was invited to participate by Wang Yiwei, who addressed the theme with a rhetorical question: Tea and coffee are both enjoyable, so why do we not mix them? It was a fitting metaphor for BRI's uneasy relationship with extant international alliances, which Evandro Carvalho of Brazil's Getúlio Vargas Foundation discussed with me after the recording:

> A key factor against Brazil joining BRI is US pressure. There are arguments on both sides because maybe BRI wouldn't add much economically, but on the other hand, it could make Brazil more attractive for investment. This topic isn't openly discussed in Brazilian politics, which is strange given its importance. I think the right doesn't want this to be debated because they're happy with the status quo, like I told you about agriculture. Only our former president, Celso Amorim, who now advises Lula, openly says he thinks we should join BRI. I was astonished that BRI wasn't even mentioned in the formal proceedings of Lula's recent visit. . . . US companies are closely tied to the Brazilian right, and I don't see how can BRI can overcome this. (Interview October 19, 2023)

Evandro's observation—that Brazilian public debates about China's growing global influence are carefully managed—holds true for other contexts, including China and the United States. This is apparent in the edited version of the talk show that was aired a few days later and in a comparable experience I had with a Washington, DC, think tank, described in the next section. Participating in predestined "debates" raises ethical dilemmas for Chinese intermediaries and foreign researchers who work with them. For the former, conformity to official narratives is the price of secure professional status, constituting a core proviso for their connective work. For the latter, a fundamental question arises: how to convey mutually constructive ideas between distinct—even opposing—groups without falling afoul of both?

Ethical Speedbumps

BRI has generated a saying among Chinese academics: It is an ideal thesis topic for students you love and for students you hate. Both will politely thank you because there is so much literature for them to use, and both will quietly curse you because there is so little they can freely write. The saying applies not only to Chinese students but also to intermediaries, who creatively reframe a wide range of Chinese traditions to explain BRI but must conform to official narratives about its purpose. This paradox of conformity and creativity reflects the tension between inward bonding and outward bridging that permeates their work. It is evident in the simmering ambivalence about the *nei can* process expressed by my host in the anecdote at the chapter's outset.

The nascent malaise about policy work that I detected among my Chinese colleagues prompted me to investigate further, as I perceived their role in influencing public opinion and political process both impressive and troubling. Like many researchers, over the years I have accepted commissions to write policy briefs for think tanks, enterprises, and governments, but the *nei can* process seemed to represent a more streamlined channel of critical engagement. What lessons, I wondered, does this process hold, and would my Chinese colleagues recommend such a system? Posing these questions seemed to me a constructive way to understand the formulation of Chinese policies, not least toward agriculture suppliers. I expected the resulting conversations to raise delicate issues, but the ethical dilemmas and insights they uncovered went deeper than I imagined. Over the past decade I have been surprised by the disposition and even enthusiasm of my Chinese colleagues to discuss such sensitive matters, which most say they are rarely asked to reflect on. I have removed their names and randomized their genders in the following interviews, together with dates and any other identifying details.

In the opening anecdote, my host indicated that writing a *nei can* is "a detailed process and not for open publication—even the author's name is removed." He acknowledged the merits of collective advancement, but like the following researcher, lamented the lack of personal acclaim:

> I wrote a *nei can* suggesting that the BRICS should accept more members. A few months ago, the BRICS did exactly this by accepting six new members at the

> summit in Johannesburg. Did the Chinese government follow my suggestion? I don't know and I never will. The leaders know if the advice they get is good or not, but it's very hard for an author to know if their advice was followed. This is a problem because people here are so competitive, and we all need to get our work recognized. . . . It's a lot of work because in universities and think tanks we're expected to write academic articles and books, but we're encouraged more and more to write policy papers that we don't get much personal credit for.

Colleagues in Shanghai reported a higher degree of intellectual choice than in Beijing, a difference that one explained as historical:

> It's a question of innovation. If you analyze only what the government tells you, then you will be limited in your own thoughts. Beijing is the center of government and people look inward, while here we are historically more international and outward looking.

Such assertions of personal choice were rare in Beijing, where being "encouraged" to write internal policy papers was increasingly common. As one Beijing scholar put it: "Getting dragged into policy advice is the last thing I want, but more and more we are expected to do it. Imagine having to write more and more advice for someone who never tells you if they read it!" Another perceived a correlation of personal sacrifice and professional status:

> There is such high competition that many people want to show that their work is more than simply academic. You have to link your academic work with your policy papers by talking about the Silk Road and other traditions from history and, actually, this becomes expected of you when your ideas get more attention. . . . Even though your name is removed, people in the system know your contributions and you can get promoted up because of them. But the higher you go, the more you have to sacrifice yourself and your family for the common good. You're always on call to answer questions, even in the middle of the night. The more policy work you do, the more you learn and the more useful your ideas get; but the more useful you get, the more careful you have to be what you write and say.

Caution about "what you write and say" was evident in an encounter at a Beijing think tank. Six agriculture specialists, most of whom I had known for years, sat opposite me in suits and ties behind name cards as their director introduced them in order of rank. My questions about Chinese urbanization and its consequences for foreign commodity exporters were answered with routine statements about "mutual benefits" and increasing trade statistics that I could have found online. After the meeting the director invited me to join the team for lunch, and as we stepped through the traditional gate marking the institute's boundary, the meeting's formality dissolved. The director set the tone with a candid remark:

> You know, Adrian, every day I walk home past the Zhongnanhai [government leadership district], but not once have I been inside. They never ask us for advice because their policies are already decided, and they are too busy to read the opinions of so many people. But still, it is our job to advise them and write *nei can*. That's how we get promoted and how we show that we understand all the policies and their history, and that we support them.

The director's experience aligns with the observation of other researchers that conformity to prevailing policies and their historical narratives can elevate professional status. While job security is a defining proviso for all of the intermediaries discussed in this book, achieving it in China requires what one foreign affairs specialist called "demonstration of loyalty." She referred to "common knowledge" that overt criticism of existing policies would be redacted by supervisors and might impede professional advancement.

I suggested to several researchers that the *nei can* system seems to carry great merit as a *form* of communication, and yet restrictions on *content* may undermine this merit. Most disagreed with this point, suggesting instead that "you can say anything if you know how to say it," and that "not being too harsh" is an important skill. More than a decade ago, Peking University international relations scholar Professor Zha Daojiong expressed a similar point:

> I think the trick is to say things in a way that is nonconfrontational, that is, make comments to let the policymakers know what issues should be more in the foreground. What we don't do, though, is what the Indian and Philippine analysts do: openly publish our hunches and whims as though they are fact. Some things just shouldn't be published because it is better to consider them internally, and that means being diplomatic. (Interview, September 27, 2012)

An example of diplomacy was offered by a food security analyst, who recounted how scholars advised the government during COVID:

> During the pandemic we had long lockdowns and there were many people who advised the government to make sure the public has enough fresh food. To present this idea in a way that was convincing and not controversial, they made it relate to stimulating economic growth. So, you see? They made sure that their advice supports the government's priorities. . . . Writing a *nei can* is hard work because first you have to know with a lot of detail about many different policies, and then your advice has to help the success of these policies.

Supporting existing policies is a requirement for Chinese advisors, but this poses challenges for those who are called to mediate with foreign political systems. One researcher struggled to convey mistaken assumptions to a Chinese enterprise, which had suffered delays when the Brazilian government failed to implement an agriculture agreement:

> Our system in China gives the government the authority to push through reforms. We say it has the power to "grip with the hand" [抓手, *zhuā shǒu*] to directly make things happen. But in Brazil, the national government doesn't have this ability because local governments and communities often disagree with its decisions, so we should not assume they will accept our way of doing things. This has caused some mistakes by Chinese enterprises and politicians, but it is difficult to say this directly, so I should say these are not mistakes, and actually they are cultural misunderstandings from different civilizations.

The researcher's sensitive approach subsequently generated a request to write a full *nei can*, whose contents he planned to gather while attending an academic conference in Brazil. Like several others I spoke with, he viewed university seminars and symposia as good opportunities for collecting data relevant to Chinese policy challenges. Two other scholars were planning trips to Brazil and Argentina to participate in conferences, and both intended to combine these with side meetings to inform *nei can* papers. Another described a similar approach to writing a policy report on tourism:

> I received an invitation from the Ministry of Culture to write a report on tourism in South America, so it paid for me to travel so I could go and learn about this. I went and gave some talks and conferences there and in the USA, and then the questions and things people said showed me what they think. . . . You know the Chinese government has a long perspective, so I learned about the Chinese workers who went there in the 1800s. That's important for helping Chinese tourists to feel connected and welcome. . . . I can do this and move freely because I'm a scholar and not a politician.

Political neutrality is assumed of scholars, but not all scholars are politically neutral. One North American researcher I met in Beijing was a strident critic of US foreign policy and had relocated to China to study BRI's pursuit of international "modernization." She spoke brazenly about President Xi's global vision as a superior alternative to prevailing liberal democratic norms, expressing a zero-sum perspective not generally shared by Chinese scholars. Her statements afforded little room for the notions of collaboration and "harmony" that permeate Xi's discourse, foregrounding instead the incompatibility of the Chinese and US systems. Uninterested in common interests and "win-win" cooperation, her focus on irreconcilable differences resembled recent calls from conservative Western think tanks to counter China's rise by "preparing for great power conflict" and "winning the new cold war" through "impactful projection" and other adversarial tactics (Carafano et al. 2023; Cozad et al. 2023; Hellyer and Nicholls 2022).

Strategic concerns are a valid component of foreign policymaking, but they are often singled out and sensationalized by alarmist news reports and online media. Academic researchers can contribute balanced and critically informed voices to such polarized discussions, but political neutrality is an elusive quality when the conclusions of debates are predetermined by their hosts. My attempts to offer politically neutral input into polarized discussions, recorded in my research diary, suffered a similar fate in Beijing and Washington, DC:

Mediating the Media

The talk show about BRI and modernization that was filmed in Beijing has just been aired and published online (CGTN 2023). It has been edited for time and, it seems, for content. Before the cameras rolled and as the lights settled on a purple hue, I spoke with the host to get a sense of his interests. He told me that "we try to build a sense of national unity," which I supposed meant emphasizing patriotism, conformity, and confidence in authority. It wasn't the first time I've asked myself how to say things without my words being repurposed by video editors. But these editors have done their job well: My suggestion that BRI should more explicitly promote ethnic inclusion and women's leadership has been transformed into praise for BRI's achievements in these respects. And gone is my observation that "other countries came to recognize the mistakes of modernity and tried to go postmodern." I shouldn't be surprised: This is what happens when the media tries to "build a sense of national unity."

I'm reminded of an interview on China–Cuba relations hosted by a Washington, DC, think tank earlier this year. The topic was in the news owing to allegations by the *Wall Street Journal*, amplified by CNN, of a "secret agreement for China to establish an electronic eavesdropping facility on the island, in a brash new geopolitical challenge by Beijing to the U.S." (Strobel and Lubold 2023). The White House had just stated that there was "no evidence" to substantiate the media reports, and the Pentagon said they were "inaccurate," so I suggested we take a more discerning approach to mass media reports. Although I quoted the White House and Pentagon refutations, the event summary leaves the "intelligence outpost" allegation unquestioned (NCUSCR 2023). As in the Chinese talk show, a predetermined position remained resistant to change.

Years ago I complained to a mentor, Professor Margaret Crahan, about being misquoted in the media. She offered tactical advice: "It's just part of the game; careful what you say."

Research diary, Melbourne, October 23, 2023

Figure 5.1
Careful what you say: managing the media in Beijing. Photo courtesy of the author.

Cooption by predetermined agendas, as Crahan points out, is a common problem for anybody who comments publicly on politically divisive issues. Polarized debates about China's growing global influence fortify the inward bonds of opposing sides, leaving little middle ground for building bridges. As demands for loyalty grow, intermediaries cannot easily maintain arm's length from the groups they engage and leverage the resulting "strength of weak ties" (Granovetter 1973).

The Chinese intermediaries I interviewed were acutely aware that balancing national loyalty with personal voice is a skill. Their testimonies reveal that researching Latin American responses to China's growing influence, and communicating these responses to their government, requires diplomacy. This sensitivity is evident in their discursive tactics, which include aligning *nei can* reports with existing policy priorities, making suggestions "in a way that is nonconfrontational," and "not being too harsh" when critiquing "our way of doing things." Diplomacy also entails references to the Silk Road, overseas Chinese communities, and other historical narratives that support the government's "long perspective." These tactics permit Chinese intermediaries to "say anything if you know how to say it," opening a middle ground between political loyalty and personal integrity.

Like their Brazilian, Australian, and Cuban counterparts in earlier chapters, Chinese intermediaries specialize in narratives that strengthen internal bonds by linking historical sensibilities with emerging agendas. But more than their foreign counterparts, they formulate these narratives as a condition of job security and professional status. While some complain about being "dragged into policy advice," most accept this work as an unavoidable aspect of career advancement amid intense competition.

Seeking material for a potential *nei can*, one researcher I had interviewed contacted me shortly before I left China. Soybean cultivation, he said, is a three-thousand-year-old Chinese tradition, and an agriculture enterprise he was working with in Brazil was developing a "people-to-people" project to raise awareness of this shared cultural heritage. He suggested a collaboration with my urban farming project in Rio de Janeiro (Hearn 2023), and the possibility of writing a *nei can* together about community engagement strategies. I thanked him for the proposition but also expressed concern about maintaining my project's political neutrality. As this book goes to press, that discussion continues, generating ongoing reflection on the ethical challenges that permeate the work of intermediaries.

The pursuit of engaged research is both a personal dilemma and a global challenge. As the middle ground disappears between polarized assessments of China's rise, it is also disappearing between the world's commodity plantations and its expanding cities. But creeping commodity frontiers, intertwined with shifting geopolitical boundaries, have not done away with food system intermediaries. Rather, the book's conclusion argues that this emerging confluence of local and global pressures is stimulating intermediaries to cultivate new alliances from the ground up.

Conclusion: Cultivating Middle Ground

The community farms springing up across Latin America, Australia, and China are deceptively simple. Their vibrant colors animate gray apartment blocks and congested streets with hope and vitality, but their stories reveal deep tensions between local identities and global transformation. Embodying traditions of land and livelihood, urban orchards and family farms demonstrate that the seemingly unstoppable forces of globalization rarely go uncontested. This book has found that food system intermediaries refashion these traditions to build alliances and assert local voices.

Community voices are easily drowned out by the economic success of export agribusiness. Brazilian soybeans and Australian wheat demonstrate this success and echo the story of Cuban sugar: Industrial expansion is generating macro-economic benefits but incurring micro-level costs. Among these costs are the replacement of peri-urban farms by sprawling plantations, loss of biodiversity, and contamination of groundwater with pesticides and fertilizers. As rural towns are engulfed by mechanized industries that offer few jobs and social ties, young people are seeking their future in cities. Expanding urban populations and the new suburbs that house them put further pressure on the peri-urban zones that have historically grown fresh food for local consumption.

Most rural–urban migrants possess scarce economic resources but abundant knowledge of food production traditions and techniques. Their lack of financial capital can be offset by this reserve of human capital when they become connected with community food projects. The book's case studies show that intermediaries are instrumental in this process, for it is they who broker the social capital needed for new alliances. Linking rural migrants with urban farms, government policies, and market opportunities requires middle ground, which intermediaries cultivate by articulating broadly familiar food traditions. Urban agriculture, environmental stewardship, *guanxi*, and food sovereignty are among the traditions articulated by intermediaries in this book to tap into local sensibilities while resonating across social divides.

Identifying and amplifying shared priorities is becoming more difficult as middle ground disappears between cities and commodity plantations and between opposing political perspectives. Researchers in the early years of the twenty-first century observed this widening rift, the diverging responses it provoked among food activists, and the need to overcome resulting divides. "Changing the food regime," wrote Eric Holt-Giménez in Hope Alkon and Agyeman's seminal volume, "will depend in large part on the ability of food activists to link issues and build bridges between the progressive and radical trends of the growing food movement" (2011, 327). Fifteen years on, opinions on the global food system have become even more polarized and in need of bridges as the ripples of Chinese demand extend around the world. China's global rise, underpinned by an ambitious urbanization agenda, has made it the world's leading importer of commodities, generating concerns not only about resulting social and environmental impacts but also about threats to geopolitical order. In response to foreign criticism, some Chinese enterprises are adopting a greater focus on profit-oriented activities to diminish concerns about their political agendas, while others are promising to improve on the track record of earlier colonial powers. The latter approach is reflected in the Third Belt and Road Forum's commitment to more socially engaged investment projects, but to achieve this will require intermediaries capable of cultivating physical and political middle ground.

The book's introduction proposed analyzing food system intermediaries by building on previous analyses of social capital, structural holes, and "the strength of weak ties." I then suggested that these concepts can be enriched with a culturally oriented approach that clarifies not only *how* networks form but also *why* they do so. The chapters observed intermediaries invoking food traditions to strengthen motivational bonds within distinct groups, and then emphasizing shared sensibilities to build bridges between them. Case studies showed that this culturally contingent process plays out differently around the world, but also that comparable conditions enable and impede it. Supportive policy environments, public familiarity with food traditions, and secure professional status are fundamental provisos that intermediaries simultaneously accommodate and consolidate to carry out their work.

Chapter 1 considered the shrinking spaces between community farms and agribusiness plantations, where expanding "commodity frontiers" set the stage for food system intermediaries. Brazil's soybean industry has attracted the interest of governments and corporations around the world that wish to deepen trade relations with China, but this trade is generating problems on both sides of the Pacific not seen with mining, gas, oil, and other extractive activities. Supply-side objections include overreliance on commodity exports that add little value and generate few jobs, the loss of crop diversity in rural and peri-urban zones, environmental damage, and unsustainable urbanization. These concerns are met with demand-side suspicion of safety standards, apprehension about unhealthy processed food, and personal disconnection with the productive process. As rural and urban boundaries blur, intermediaries have emerged to articulate shared purpose amid changing food systems and polarized debates about China's impact.

Chapter 2 outlined the colonial systems, premised on commodity exports to Europe, that set the Australian and Brazilian agriculture sectors in motion. Their respective reliance on wheat and soybeans, now destined for China, suggest that extractive models remain intact. Mechanization and ecological degradation have provoked rural migration to cities like Melbourne and São Paulo, whose resulting growth has compromised peri-urban productive zones and unleashed real estate developers. The latter's advertising strategies, evident in both cities, emphasize an underlying desire among prospective homeowners to reconnect with nature and with each other. But the boom of supposedly green real estate brings gentrifying effects, including "vegetation apartheid" and public distraction from the broader erosion of land, employment, and environment. Food system intermediaries pushed back against these pressures: Chris Williams planted food crops in public spaces to celebrate Dandenong's historical multiculturalism, Dona Sebastiana marshalled her reputation as a traditional farmer to legally defend her land from a hostile takeover, and activists in both cities supported peri-urban producers by linking them with online inner-city customers. These digital platforms were efficient but lacked the cultural sensitivity of human intermediaries in response daily challenges, from broken down trucks to personal negotiation with customers. The chapter concluded that intermediaries who develop narratives informed by local history more effectively attract support from municipal governments and customers, whose alliance can stimulate policy reforms and limit corporate power. But it also revealed the professional precarity that intermediaries endure, which became clear when COVID lockdowns closed restaurants, schools, and famers markets, putting many out of work. Lacking formal standing and stable jobs, some tried to continue operating on the fringes of established occupations while others chose—or were forced—to retire.

Chapter 3 visited Shared Harvest, an enterprise developed by Shi Yan in Beijing, and the Hortas Cariocas initiative led by Julio Barros and Fabiola Corrêa in Rio de Janeiro. Both projects harnessed cultivation traditions handed down over generations in commercially oriented urban farms, accommodating people displaced from rural towns into expanding small businesses. Shi's New Farmer program forged what she called "concentric rings" of CSA (community-supported agriculture) practitioners, whose cooperation was underpinned by longstanding *guanxi* relationships between rural families and their departed relatives. Julio and Fabiola articulated the historical benefits of urban farming and "ancestral plants" to encourage community participation and build bridges between disadvantaged Manguinhos residents and their surrounding suburbs. As well as strengthening public awareness of food traditions, they secured local government support by integrating their projects with emerging sustainability and health policies. While this consolidated the political and cultural provisos for their work, none of the three escaped professional precarity. Collective dependence on their personal abilities, deepened by the risk of funding cuts and policy shifts, made them and their projects valuable yet vulnerable. Conscious of this paradox, the three intermediaries identified the need to build resilience by recruiting new leaders and creating more democratic leadership structures.

Chapter 4 considered the intersection of global politics and local survival in Cuba's food system. To endure the Soviet Union's collapse, Cuba was forced to abandon centuries of export agriculture and instead feed citizens through neighborhood horticulture. Chinese support in the early 2000s and Fidel Castro's resignation in 2008 ushered in economic reforms that transformed Cuban diets once more. My friend Miki demonstrated this trend, rejecting fresh vegetables from government subsistence farms and embracing fast food served up by a nascent private sector. Cuba's 2021 food sovereignty policy pledges a return to locally grown healthy diets, coinciding with the promotion of foreign study tours that venerate the urban farming innovations of the 1990s. The push for "agroecotourism" identifies food sovereignty as a hard-won Cuban tradition, which Luis and Reinaldo articulated to navigate the restrictions and allowances of top-down policies, construct networks around a unifying theme, and pursue professional opportunities. As a locally conceived yet globally familiar aspiration, food sovereignty enabled them to advance distinct agendas, from a "circle of benefits" for neighbors to a vision of reform for the nation. The national autonomy implied by food sovereignty remains elusive as the US trade embargo reinforces Cuban reliance on Chinese investment in the agriculture and energy sectors. Under the umbrella of the Belt and Road Initiative (BRI), these activities aim to revive Cuba's export sugar industry while building capacity in corn and rice processing. The resulting paradox of what the chapter called "sovereign dependence" evokes deep-set suspicions of neocolonialism for Cuba and other commodity exporters, which the Chinese government is determined to overcome.

China has developed the BRI as a platform to engage trade and investment partners with the promise of mutual benefit. Scholars working at BRI research centers travel the world to explain the initiative and observe the reactions it is provoking in academic conferences, government summits, and corporate boardrooms. Chapter 5 showed that they are increasingly expected to fashion policy advice from these observations, and to communicate their suggestions to political leaders in *nei can* briefing papers. Balancing deference to existing policies with critical input requires Chinese intermediaries to navigate a diplomatic minefield. Ethical and practical dilemmas become acute when critiquing official assumptions, such as erroneous policy expectations among Chinese agriculture enterprises in Brazil. Amid intense professional competition, broadly familiar Chinese traditions furnish middle ground between political loyalty, personal integrity, and job security. Just as intermediaries in previous chapters deployed narratives about urban farming and food sovereignty to build consensus, their Chinese counterparts emphasized the Silk Road and other vestiges of national history to build unity and reconcile political allegiance with personal voice. This discursive tactic establishes historical precedent for China's leadership of contemporary globalization and, as one researcher explained, enables Chinese intermediaries to "say anything if you know how to say it."

The dilemmas faced by Chinese intermediaries extend to foreign researchers who critically engage with their activities. It is becoming harder to study sensitive aspects of China's rise, including the transformation of the world's food systems, without being drawn into predetermined debates. Maintaining arm's length from political agendas is a hallmark of academic integrity, yet those who investigate this momentous and controversial topic face growing pressure to take sides. Under these uncompromising conditions, the third proviso for effective intermediaries—secure professional status—cannot be assumed. Junior researchers seeking promotion, tenure, and funding are especially vulnerable to the perils that permeate the interface of Chinese globalization, community agency, and engaged scholarship. When the demise of the Soviet Union marked an earlier turning point in world history, Renato Rosaldo (1989, 26–30) argued that studying emerging intercultural "borderlands" is conceptually fascinating and politically necessary, but personally risky. Amid polarized assessments of China's engagement with Latin America and the world, how can researchers who embrace the resulting challenges and opportunities find secure middle ground?

Toward an Anthropology of the Middle

More than a century ago the German sociologist Max Weber observed that charismatic leaders, from prophets to politicians, rely on "authority from traditional custom" (1991 [1921], 249). Weber's notion that traditions provide a unifying framework for collective action foreshadowed later investigations into the ways traditions are personally adapted, collectively deployed, and strategically invented (Bourdieu 1977; Hobsbawm and Ranger 1992; Ortner 1991; Sahlins 1981). Weber also anticipated research into the mechanics of collective action through the interplay of structure and agency, the "strength of weak ties," and the divisive effects of "negative social capital" (Burt 1992; Granovetter 1973; Portes 1998; Sanjek 1974). The book's introduction outlined how these early studies into the dynamics of network formation gave rise to Robert Putnam's (2000) distinction of "bonding" and "bridging" social capital. In the twenty-first century, this dichotomy has become a prominent analytic tool that is employed to explain everything from ethnic entrepreneurs to racially diverse childcare centers.

Putnam's (2007) remark that social scientists have reduced his bonding-bridging distinction to a zero-sum pair should hinder its unreflective use. Perhaps for this reason, his remark is seldom cited. Yet the prospect of mutually reinforcing bonds and bridges opens rich middle ground between socially and culturally oriented network studies. This book has explored the gray area between inward solidarity and outward affinity by considering the role of intermediaries in building collective projects. The introduction noted that food system researchers are conceptualizing gray areas between small and large producers as the "agriculture of the middle," and suggested that their work might inspire ethnographers to develop an "anthropology of the middle."

Considering how and why intermediaries, ethnographers among them, cultivate middle ground between local identities and global transformations opens a fresh approach to an old question: How do social groups reconcile the tension between inward solidarity and outward collaboration? The book has responded to this question by suggesting that intermediaries first amplify internal bonds by associating them with broadly familiar traditions and then build external bridges by emphasizing shared values. Accommodating and consolidating the three provisos—policies, traditions, and employment—to this end, their work generates fruitful questions for an anthropology of the middle:

- How do policy conditions shape the prospects for applied anthropology, and conversely, how might ethnographers influence these conditions? New opportunities are arising from the decentralization of development funding inside (and away from) large bureaus and agencies to better support field offices, overseas embassies, and niche NGOs. Sensitive to conditions on the ground, ethnographers are well placed to identify junctures of budget delegation within and between institutions—from development banks to community associations—to inform their practices. A reflexive anthropology of the middle would treat this process as an object of study by codesigning projects at the grassroots, communicating results in ways that resonate with sponsors, and proposing improvements to their funding models.
- How might intermediaries, including ethnographers, build awareness about the capacity of food (and other) traditions to empower community agendas? An anthropology of the middle would encourage sensitivity to the subtle presence of such background cultural knowledge while reframing it to illuminate practical potentials. The intermediaries in this book harnessed sensibilities about food to advance their projects, maintaining arm's length from the groups they brought together to leverage the "strength of weak ties." Similarly, ethnographers can leverage their liminality to bring background sensibilities to the forefront of development collaboration.

- What professional conditions would enable university researchers and their students to effectively codesign community outreach and engagement projects? While academic institutions are becoming more supportive of such work, promotion and tenure panels continue to prioritize quantifiable achievements such as publications, grants, and teaching evaluations. A "middle" perspective identifies mutual benefits, including academic writing about case studies, funding from development agencies, and educative encounters between students and partner communities. Ethnographers and their employers can help each other to open these pathways, which would encourage the emergence of academic intermediaries across disciplines and career stages.
- Under increasingly polarized political conditions, what strategies can ethnographers adopt to contribute balanced perspectives? For those engaged with contentious topics, such as those explored in this book, collaboration with counterparts in China and the United States is a delicate endeavor. To promote dialogue while maintaining academic integrity, researchers are challenged to avoid cooption into predetermined debates, design seminars and symposia that promote communication across divides, and work with publishers that favor diverse perspectives.
- What can an anthropology of the middle contribute to contested scenarios beyond the transformation of food systems? Countless communities are divided by conflicting interpretations of historical identity, from the Yasukuni shrine in Tokyo to the Al Aqsa Mosque in Jerusalem and *Terra Nullius* in Australia. The book has suggested that the inward bonds treasured by distinct groups are embedded in sensibilities that can resonate outward across divides. While traditions of family, ecology, and even the divine are held dear in unique ways, a "middle" outlook perceives the capacity of individuals and groups to see themselves in each other.

These lines of inquiry envision researchers fruitfully studying and reflexively becoming intermediaries. Developing this critically engaged vision is a valuable pursuit as global forces and community agendas intersect in new ways. The consequences of this intersection for food systems are clear in the contested green zones caught between expanding commodity plantations and sprawling suburbs. Researchers of the "middle" will find these fertile borderlands rich with opportunities to contemplate resistance and cultivate alliances.

References

ABRASCO (Associação Brasileira de Saúde Coletiva). 2016. "Nova ferramenta de dados sobre agrotóxicos no Brasil reforça produção científica no tema." http://www.abrasco.org.br/site/noticias/saude-da-populacao/nova-ferramenta-de-monitoramento-de-dados-sobre-agrotoxicos-no-brasil-reforca-producao-cientifica-no-tema/20159/.

Acanda, Jorge Luis. 1997. "Releyendo a Gramsci: Hegemonía y Sociedad Civil." *Temas* 11: 75–86.

ACFS (Australia-Cuban Friendship Society). 2019. Southern Cross Brigade. http://cubabrigade.org.au/.

Adler, Patricia A., and Peter H. Adler. 1987. *Membership Roles in Field Research*. Newbury Park, CA: Sage Publications.

Alejandro, Carlos F. Díaz. 1985. "Argentina, Australia and Brazil Before 1929." In *Argentina, Australia and Canada Studies in Comparative Development 1870–1965*, edited by D. C. M. Platt and Guido di Tella, 95–109. London: Palgrave Macmillan.

Altieri, Miguel, and Fernando R. Funes Monzote. 2012. "The Paradox of Cuban Agriculture." *Monthly Review* 63 (8): 22–33.

Altieri, Miguel, Nelso Companioni, Kristina Cañizares, Catherine Murphy, Peter Rosset, Martin Bourque, and Clara I. Nicholls. 1999. "The Greening of the 'Barrios': Urban Agriculture for Food Security in Cuba." *Agriculture and Human Values* 16: 131–140.

Altruvistas. 2019. Cuba Agroecology Tour. https://altruvistas.com/cuba-agroecology-tour-ogs-january-2021/.

Amato-Lourenco, Luís Fernando. 2018. *Agricultura urbana: Guia de boas práticas*. São Paulo: USP Instituto de Estudos Avançados. http://www.iea.usp.br/pesquisa/grupos-de-estudo/grupo-de-estudos-de-agricultura-urbana/publicacoes/cartilhasiteiea.pdf/view.

Amato-Lourenco, Luís Fernando, Rafael Junqueira Buralli, Guilherme Reis Ranieri, Adrian H. Hearn, Chris Williams, and Thais Mauad. 2020. "Building Knowledge in Urban Agriculture: The Challenges of Local Food Production in São Paulo and Melbourne." *Environment, Development and Sustainability* 23: 2785–2796.

Amato-Lourenco, Luís Fernando, Tiana Carla Lopes Moreira, Vanessa Cristina de Oliveira Souza, Fernando Barbosa Jr., Mitiko Saik, Paulo Hilário Nascimento Saldiva, and Thais Mauad. 2016. "The Influence of Atmospheric Particles on the Elemental Content of Vegetables in Urban Gardens of Sao Paulo, Brazil." *Environmental Pollution* 216: 125–134.

Anderson, Molly D. 2024. *Transforming Food Systems: Narratives of Power*. London: Routledge.

Anguelovski, Isabelle, and James Connolly. 2018. "From Landscapes of Utopia to the Margins of the Green Urban Life." *City* 22 (3): 417–436.

Armony, Ariel. 2011. "The China-Latin America Relationship: Convergences and Divergences." In *China Engages Latin America: Tracing the Trajectory*, edited by Adrian H. Hearn and José Luis León-Manríquez, 23–50. Boulder, CO: Lynne Rienner.

Bachelard, Michael. 2018. "Readers Respond: Should Australia Support China's Belt and Road Initiative." *The Sydney Morning Herald*, June 26. http://www.smh.com.au/world/asia/readers-respond-should-australia-support-china-s-belt-and-road-initiative-20180625-p4znna.html.

Backhouse, Megan. 2016. "The Man on a Mission to Broaden Our Tastes." *The Sydney Morning Herald*, April 22. http://www.smh.com.au/entertainment/the-man-on-a-mission-to-broaden-our-tastes-20160422-gocxsf.html.

Bai, Tongdong. 2023. "Confucianism in Mainland China." In *The Oxford Handbook of Confucianism*, edited by Jennifer Oldstone-Moore, 230-C17P47. Oxford: Oxford Academic.

Baker McKenzie. 2017. *Belt and Road: Opportunity and Risk*. http://www.bakermckenzie.com/en/insight/publications/2017/10/chinas-belt-road-initiative.

Bampton, Bill. 2010. "H.V. McKay Gardens, Sunshine: An Industrial Garden 100 Years On." *Australian Garden History* 21 (3): 10–15.

Baptista da Costa, Manoel B., Monique Souza, Vilmar Müller Júnior, Jucinei José Comin, and Paulo Emílio Lovato. 2017. "Agroecology Development in Brazil between 1970 and 2015." *Agroecology and Sustainable Food Systems* 41 (3–4): 276–295.

Barbon, Júlia, and Ian Cheibub. 2021. "Horta gigante complementa refeição de 800 famílias em favela do Rio na pandemia." *Folha de S. Paulo*, April 24.

Barboza, David. 2008. "Chinese destroy 3,600 tons of tainted feed: Officials destroy tons of melamine-laced feed for livestock." *New York Times*, November 3.

Barthes, Roland. 1957. *Mythologies*. Paris: Seuil.

Bastian, Hope. 2019. *Everyday Adjustments in Havana: Economic Reforms, Mobility, and Emerging Inequalities*. Lanham, MD: Rowman & Littlefield.

Bastian, Hope, and Hanna Garth. 2020. "Cuban Food Security in a Time of COVID-19." *Anthropology News*, September 25. https://doi.org/10.14506/AN.1507.

Bax, Mart. 1978. Figuration Analysis: A Better Perspective for the Networker. *Anthropological Quarterly* 51 (4): 221–230.

BBC (British Broadcasting Corporation). 1988. "Egg Industry Fury over Salmonella Claim." *BBC News*, December 10. http://news.bbc.co.uk/onthisday/hi/dates/stories/december/3/newsid_2519000/2519451.stm.

Behar, Ruth. 1990. "Rage and Redemption: Reading the Life Story of a Mexican Marketing Woman." *Feminist Studies* 16 (2): 223–258.

Benítez Rojo, Antonio. 1992. *The Repeating Island: The Caribbean and the Postmodern Perspective*. Durham: Duke University Press.

Benjamin, Medea, Joseph Collins, and Michael Scott. 1989. *No Free Lunch: Food & Revolution in Cuba Today*. San Francisco: Food First.

Berardo, Ramiro. 2014. "Bridging and Bonding Capital in Two-Mode Collaboration Networks." *Policy Studies Journal* 42 (2): 197–225.

Biazoti, André Ruoppolo, Angélica Campos Nakamura, Gustavo Nagib, Vitória Oliveira Pereira de Souza Leão, Giulia Giacchè, and Thais Mauad. 2021. "The Impact of COVID-19 on Urban Agriculture in São Paulo, Brazil." *Sustainability* 13(11): 6185. https://doi.org/10.3390/su13116185.

Bita, Natasha. 2011. "Land Rush." *The Australian*, July 2.

Blaj, Ilana. 1998. "Agricultores e comerciantes em São Paulo nos inícios do século XVIII: O processo de sedimentação da elite paulistana." *Revista brasileira de História* 18 (36): 281–296.

Bloomberg Foundation. 2016. "São Paulo: Growing Farmers' Income, Shrinking Urban Sprawl." Bloomberg Philanthropies Mayor's Challenge. https://mayorschallenge.bloomberg.org/ideas/sao-paulo/.

Bloomberg Foundation. 2017. "São Paulo launches 'Connect the Dots' Initiative." https://medium.com/@BloombergCities/s%C3%A3o-paulo-launches-connect-the-dots-initiative-ddf751fa2f35.

Boas Noticias. 2022. "Brasil: Eu poderia estar morta—A horta que salva vidas." Editorial, *Boas Noticias*, May 30. https://boas-noticias.com/brasil-eu-poderia-estar-morta-a-horta-que-salva-vidas/.

Böhme, Michaela. 2021. "Shared Interest or Strategic Threat? A Critical Investigation of Political Debates and Regulatory Responses to Chinese Agricultural Investment in Australia." *Globalizations* 18 (3): 441–460.

Bourdieu, Pierre. 1977. *Outline of a Theory of Practice*. New York: Cambridge University Press.

Bridge, Gary, Tim Butler, and Patrick Le Gals. 2014. "Power Relations and Social Mix in Metropolitan Neighbourhoods in North America and Europe; Moving beyond Gentrification?" *International Journal of Urban and Regional Research* 38 (4): 1133–1141.

Brislen, Lilian. 2018. "Meeting in the Middle: Scaling-up and Scaling-over in Alternative Food Networks." *Culture, Agriculture, Food and Environment* 40 (2): 105–113.

Brizuela, Luis. 2024. "Cuba aspira a reavivar su industria cañera más allá del azúcar." *Nodal*, May 7. https://www.nodal.am/2024/05/cuba-aspira-a-reavivar-su-industria-canera-mas-alla-del-azucar/.

Brotherton, P. Sean. 2008. "We Have to Think like Capitalists but Continue Being Socialists: Medicalized Subjectivities, Emergent Capital, and Socialist Entrepreneurs in Post-Soviet Cuba." *American Ethnologist* 35 (2): 259–274.

Burns, Catherine Mary, and A. Dave Inglis. 2007. "Measuring Food Access in Melbourne: Access to Healthy and Fast Foods by Car, Bus and Foot in an Urban Municipality in Melbourne." *Health & Place* 13 (4): 877–885.

Burt, Ronald. 1992. *Structural Holes: The Social Structure of Competition.* Cambridge, MA: Harvard University Press.

Butler, Tim. 2006. "Gentrification and Globalization: The Emergence of a Middle Range Theory?" Available on Sciences Po. http://blogs.sciences-po.fr/recherche-villes/gentrification-and-globalization-the-emergence-of-a-middle-range-theory/.

C40 Cities. 2019. Food, Waste & Water Initiative. https://www.c40.org/what-we-do/scaling-up-climate-action/food-systems/

CAFS (Cuban American Friendship Society). 2020. UVM Study of Cuban Farming Methods. http://www.cubanamericanfriendshipsociety.org/post/uvm-study-of-cuban-farming-methods.

Callanan, Tim. 2021. "What is China's Belt and Road Initiative and What Were the Four Deals the Federal Government Tore up?" *ABC News*, April 22. https://www.abc.net.au/news/2021-04-22/what-was-in-victoria-belt-and-road-deal-with-china/100086224.

Camara dos Deputados. 2015. Projeto de Lei No.610 de 2015: Dispõe sobre a obrigatoriedade da inclusão de alimentos orgânicos na alimentação escolar e dá outras providências. http://ww.camara.gov.br/sileg/integras/1312230.pdf.

Campbell, Alissa M., and Rod MacRae. 2013. "Local Food Plus: The Connective Tissue in Local/sustainable Supply Chain Development." *Local Environment* 18 (5): 557–566.

Campbell, Scott. 2003. "Case Studies in Planning: Comparative Advantages and the Problem of Generalization." University of Michigan Urban and Regional Research Collaborative Working Paper URRC 02–07.

Camus, Juan Andrés, Jane LePham, Roshan Shankar, and Kenny White. 2013. *Strictly Business? An Examination of China's Natural-Resource Acquisition Strategy in Latin America*. Report for the U.S. Government Accountability Office, Stanford: Stanford University.

Capone, Alesha. 2015. "New Heart Foundation Report Reveals High Levels of Obesity and Physical Inactivity in Melbourne's West." *Brimbank Leader*, May 13.

Carafano, James J., Michael Pillsbury, Jeff M. Smith, and Andrew J. Harding. 2023. "Winning the New Cold War: A Plan for Countering China." Heritage Foundation, special report #270. https://www.heritage.org/sites/default/files/2023-07/SR270.pdf.

Cardim. 2017. "Arquitetura Paisagística Sustentável." https://www.cardimpaisagismo.com.br.

Cardoso Carrero, Gabriel, Cynthia S. Simmons, and Robert T. Walker. 2022. "The Great Amazon Land Grab—How Brazil's Government is Clearing the Way for Deforestation." *The Conversation*, February 3.

Cardoso, Fernando. 2001. *Charting a New Course: The Politics of Globalization and Social Transformation*. Lanham: Rowman & Littlefield.

Cardoso, Fernando, and Enzo Faletto. 1979. *Dependency and Development in Latin America*. Berkeley: University of California Press.

Carey, Rachel, Jennifer Sheridan, and Kirsten Larsen. 2018. *Food for Thought: Challenges and Opportunities for Farming in Melbourne's Foodbowl*. Melbourne: Victorian Eco-Innovation Lab.

Carey, Rachel, Maureen Murphy, Leila Alexandra, Jennifer Sheridan, Kirsten Larsen, and Emily McGill. 2022. *Building the Resilience of Melbourne's Food System—A Roadmap*. Melbourne: Victorian Eco-Innovation Lab.

Carrai, Maria A. 2021. "Adaptive Governance Along Chinese-Financed BRI Road Megaprojects in East Africa." *World Development* 141 (1). https://doi.org/10.1016/j.worlddev.2020.105388.

Carrier, James G. 1999. "People Who Can Be Friends: Selves and Social Relationships." In *The Anthropology of Friendship*, edited by Sandra Bell and Simon Coleman, 21–38. New York: Berg.

Castro, Fidel. 2001. Discurso pronunciado por el Comandante en Jefe Fidel Castro Ruz en el Foro Mundial sobre Soberanía Alimentaria. http://www.fidelcastro.cu/es.

Catacora-Vargas, Georgina, Pablo Galeano, Sarah Zanon Agapito, Darío Aranda, Tomás Palau, and Rubens Onofre Nodari. 2012. *Soybean Production in the Southern Cone of the Americas: Update on Land and Pesticide Use*. Cochabamba: Virmegraf.

Catcomm. 2023. Rio Favela Facts. https://catcomm.org/favela-facts/.

CBBC (China-Brazil Business Council). 2011. *Chinese Investments in Brazil: A New Phase in the China-Brazil Relationship*. Rio de Janeiro: CBBC.

CCICED (China Council for International Cooperation on Environment and Development). 2020. *Green BRI and 2030 Agenda for Sustainable Development*. Beijing: CCICED.

CEIC. 2023. Brazil Social: Poverty and Inequality. https://www.ceicdata.com/en/brazil/social-poverty-and-inequality.

CERES (Centre for Education and Research in Environmental Strategies). 2019. CERES Global: Cuba. https://ceres.org.au/global/cuba/page/6/.

CERES (Centre for Education and Research in Environmental Strategies). 2023. "Welcome to CERES." https://ceres.org.au/about/.

CGD (City of Greater Dandenong). 2014. *Greater Dandenong Regional Food Strategy 2015–2018*. Melbourne: The Field Institute.

CGD (City of Greater Dandenong). 2017. *Profile of Health and Wellbeing in Greater Dandenong 2016*. Melbourne: Dandenong City Council.

CGTN (China Global Television Network). 2023. "BRI—A New Path of Modernization for All." October 23. https://news.cgtn.com/news/2023-10-22/Live-BRI-A-new-path-of-modernization-for-all-1o6P9iwW4WQ/index.html.

Chan, May Ling, and Eduardo Francisco Freyre Roach. 2012. *Unfinished Puzzle: Cuban Agriculture: The Challenges, Lessons, and Opportunities*. Oakland, CA: Food First Books.

Chatterjee, Rhitu. 2016. "Brazil's School Lunch Program is Putting Food on the Table for the Country's Small Farmers." Public Radio International, February 10. http://www.pri.org/stories/2016-02-10/brazils-school-lunch-program-putting-food-table-countrys-small-farmers.

Chaudhuri, Pramit Pal. 2018. "Opinion: India Nudges China toward Belt and Road Changes." *Geopolitical Intelligence Services*, January 4. http://www.gisreportsonline.com/opinion-india-nudges-china-toward-belt-and-road-changes,politics,2762.html.

Chen, Mingxing, Yuan Zhou, Xinrong Huang, and Chao Ye. 2021. "The Integration of New-Type Urbanization and Rural Revitalization Strategies in China: Origin, Reality and Future Trends." *Land* 10 (2). https://doi.org/10.3390/land10020207.

Cheng, Tsung O. 2011. "Diabetes Epidemic in China and Its Economic Impact." *International Journal of Cardiology* 149 (1): 1–3.

Cheng, Yinghong. 2007. "Sino-Cuban Relations and Cuba's Future after Fidel Castro." *History Compass* 5 (2): 725–736.

Cheng, Yinghong. 2012. "The 'Socialist Other': Cuba in Chinese Ideological Debates since the 1990s." *The China Quarterly* 209: 198–216.

Chu, Yin-wah. 2020. "China's New Urbanization Plan: Progress and Structural Constraints." *Cities* 103. https://doi.org/10.1016/j.cities.2020.102736.

Chung, Stella. 2012. "Evaluating CERES Fair Food as an Urban Sustainable Food Initiative: A Comparison of Retail Pricing, Wholesale Pricing and Food Miles for a Healthy Food Basket." Independent Study Project (ISP) Collection. Paper 1269.

Clarke, Philip. 2013. "Edwina Currie: 'No Regrets on Salmonella Crisis.'" *Farmers Weekly*, December 3. https://www.fwi.co.uk/livestock/poultry/edwina-currie-no-regrets-on-salmonella-crisis.

Clifford, James, and George E. Marcus, eds. 1986. *Writing Culture: The Poetics and Politics of Ethnography*. Berkeley: University of California Press.

Collins, Samuel G., and Matthew S. Durington. 2015. *Networked Anthropology: A Primer for Ethnographers*. London: Routledge.

Comassetto, Bruno Henrique, Gabriela P. Solalinde, João Vicente Rosa de Souza, Marcelo Trevisan, Paulo Ricardo Zilio Abdala, and Carlos Alberto Vargas Rossi. 2013. "Nostalgia, anticonsumo simbólico e bem-estar: A agricultura urbana." *Revista de Administração de Empresas* 53 (4): 364–375.

Condon, Jon. 2017. "China Slaps Ban on Brazilian Exports over Processing 'Scandal.'" *Beef Central*, March 20. http://www.beefcentral.com/news/china-slaps-ban-on-brazilian-exports-over-processing-scandal/.

Coronil, Fernando. 1995. "Introduction." In Fernando Ortiz, *Cuban Counterpoint: Tobacco and Sugar*, ix–xvi. Durham: Duke University Press.

Corrêa, Jéssica Díez. 2019. "Fachadas verdes ganham novo 'contorno' em prédios." *Estadão*, February 10. https://economia.estadao.com.br/blogs/radar-imobiliario/8009-2/.

Couser, G. Thomas. 2004. *Vulnerable Subjects: Ethics and Life Writing*. Ithaca, NY: Cornell University Press.

Cozad, Mark, Keith Gierlack, Cortez A. Cooper III, Susan G. Straus, Sale Lilly, Stephanie Anne Pillion, and Kelly Elizabeth Eusebi. 2023. *Preparing for Great Power Conflict: How Experience Shapes U.S. and Chinese Military Training*. Santa Monica, CA: RAND Corporation.

CPDRC (China Population and Development Research Center). 2008. "2006 China's Urban and Rural Population by Region." **2006** 年中国各地区城镇、乡村人口数. https://web.archive.org/web/20090310163630/http://www.chinapop.gov.cn/wxzl/rkgk/200806/t20080629_157020.htm.

Crahan, Margaret E., and Ariel C. Armony. 2007. Does Civil Society Exist in Cuba? Florida International University, Cuban Research Institute. https://cri.fiu.edu/research/commissioned-reports/civil-society-crahan0armony.pdf.

Cranston, Matthew. 2012. "Chinese Want Slice of Rural Australia." *The Australian Financial Review*, January 23.

Crouch, David, and Colin Ward. 1997. *The Allotment: Its Landscape and Culture*. 2nd ed. London: Faber and Faber.

CTO (Caribbean Tourism Organization). 2002. "Tourist Arrivals by Market 2001." https://www.onecaribbean.org/statistics/historical-data-1970-2015/.

Cuba Platform. 2020. Food Sovereignty. Caring in Crisis blog series. https://cubaplatform.org/blog/2020/4/13/caring-in-crisis-food-sovereignty.

Cuban United Nations Association. 2000. *Contribución de las ONGs Cubanas e Internacionales Radicadas en Cuba Para el Forum del Milenio*. Havana: CUNA.

Cullen, Christoper. 2017. *Heavenly Numbers: Astronomy and Authority in Early Imperial China*. Oxford: Oxford University Press.

CUSAN (Cuba-US Agroecology Network). 2020. About CUSAN. http://www.agroecologynetwork.org/about.html.

Dadabaev, Timur. 2018. "'Silk Road' as Foreign Policy Discourse: The Construction of Chinese, Japanese and Korean Engagement Strategies in Central Asia." *Journal of Eurasian Studies* 9 (1): 30–41.

Datamark. 2016. "Legislation Strengthens Organic Consumption." *Datamark*, 29th February. http://www.datamark.com.br/noticias/2016/2/legislation-strengthens-organic-consumption-202917/.

Davenport, Sarah Grace, and Joanna Mishtal. 2019. "Whose Sustainability? An Analysis of a Community Farming Program's Food Justice and Environmental Sustainability Agenda." *Culture, Agriculture, Food and Environment* 41 (1): 56–65.

De Master, Kathryn. 2018. "New Inquiries into the Agri-Cultures of the Middle." *Culture, Agriculture, Food and Environment* 40 (2): 130–135.

De Paiva, Fernanda. 2022. "Jeitinho (Brazil)." Global Informality Project. http://www.in-formality.com/wiki/index.php?title=Jeitinho_(Brazil)#cite_note-1.

De Paula, Tainá. 2023. Tainá de Paula: Soberania alimentar é direito de todo cidadão brasileiro. *O Día*, June 20. https://odia.ig.com.br/opiniao/2023/06/6655412-taina-de-paula-soberania-alimentar-e-direito-de-todo-cidadao-brasileiro.html.

De Sousa, Ana Tereza L. Marra, Giorgio Romano Schutte, Rafael Almeida F. Abrão, and Valéria Lopes Ribeiro. 2023. "China in Latin America: To BRI or not to BRI: The Case of the Belt and Road Initiative." In *The Palgrave Handbook of Globalization with Chinese Characteristics*, edited by Paulo Afonso B. Duarte, Francisco José B. S. Leandro, and Enrique Martínez Galán, 495–514. London: Palgrave.

Dooling, Sarah. 2009. "Ecological Gentrification: A Research Agenda Exploring Justice in the City." *International Journal of Urban and Regional Research* 33 (3): 621–639.

Dean, Warren. 1995. *With Broadax and Firebrand: The Destruction of the Brazilian Atlantic Forest*. Berkeley: University of California Press.

DEDJTR. 2017. *Agriculture Victoria Strategy: Supporting Victoria's Agriculture*. Melbourne: Department of Economic Development, Jobs, Transport and Resources.

Descola, Philippe. 1996. "Constructing Natures: Symbolic Ecology and Social Practice." In *Nature and Society: Anthropological Perspectives*, edited by Philippe Descola and Gisili Palsson, 82–102. London: Routledge.

Diamond, Larry. 1999. *Developing Democracy: Toward Consolidation*. Baltimore: Johns Hopkins University Press.

Díaz Pérez, Maidelyn, Yuri Triana Velázquez, Pablo Brizuela Chirino, Reinaldo Javier Rodríguez Font, Raudel Giráldez Reyes, and Joovaim Blanco Borrego. 2021. "Soberanía Alimentaria y Educación Nutricional desde la ciencia de la sostenibilidad: Observatorio SAEN+C Pinar." *Revista Universidad y Sociedad* 13 (5): 9–19.

Dumont, Jean-Paul. 1978. *The Headman and I: Ambiguity and Ambivalence in the Fieldworking Experience*. Austin: University of Texas Press.

Duran, Ana Clara, Ana V. Diez Roux, Maria do Rosario D.O. Latorre, and Patricia Constante Jaime. 2013. "Neighborhood Socioeconomic Characteristics and Differences in the Availability of Healthy Food Stores and Restaurants in Sao Paulo, Brazil." *Health & Place* 23: 39–47.

Dun, Olivia, D. Bogenhuber, L. Head, J. Kadahari, N. Klocker, J. Niyera, and J. Sindayigaya. 2018. "Bringing Together Landless Farmers and Unused Farmland: The Sunraysia Burundian Garden and Food Next Door initiative." In *Reclaiming the Urban Commons: The Past, Present and Future of Food Growing in Australian Towns and Cities*, edited by Nick Rose and Andrea Gaynor, 39–52. Crawley: UWA Publishing.

Dussel Peters, Enrique, ed. 2019. *China's Foreign Direct Investment in Latin America and the Caribbean: Conditions and Challenges*. Boulder: Lynne Rienner Publishers.

Eckstein, Susan Eva. 1994. *Back from the Future: Cuba under Castro*. Princeton: Princeton University Press.

ECLAC (Economic Commission for Latin America and the Caribbean). 2021. "Cities and Housing Provide an Opportunity to Transform Latin America." *ECLAC News*, May 17. https://www.cepal.org/en/news/cities-and-housing-provide-opportunity-transform-latin-america-and-caribbeans-development-model.

Ecocity Builders. 2019. "What is an Ecocity?" https://ecocitybuilders.org/what-is-an-ecocity/#ecologicallyhealthycity

Ecoeficientes. 2019. "Sobre nós." http://www.ecoeficientes.com.br/escritorio-de-arquitetura-sustentavel-ecoeficientes/.

Economist. 2017. "São Paulo's Mayor Tries to Make the City Greener." April 27. http://www.economist.com/news/americas/21721424-planting-trees-concrete-jungle-s-o-paulos-mayor-tries-make-city-greener.

Economist. 2023. The Global Liveability Index 2023. Economist Intelligence Unit. https://www.eiu.com/n/campaigns/global-liveability-index-2023/.

Edelman, Marc, Tony Weis, Amita Baviskar, Saturnino M. Borras Jr., Eric Holt-Giménez, Deniz Kandiyoti, and Wendy Wolford. 2014. "Introduction: Critical Perspectives on Food Sovereignty." *Journal of Peasant Studies* 41 (6): 911–931.

Ehteshami, Anoushiravan, Benjamin Houghton, and Jia Liu. 2024. *China Moves West: The Evolving Strategies of the Belt and Road Initiative.* Boulder: Lynne Rienner.

Elder, John. 2023. "Barely a Pass: CSIRO Marks Down Our National Eating Habits." *The New Daily*, September 15. https://www.thenewdaily.com.au/life/2023/09/15/our-failing-national-eating-habits.

Ennis, Chris. 2012. "CERES—Once We Were Farmers." *Geodate* 25 (3): 6–8.

Eslake, Saul. 2011. "The Counterexample of Australia." In *What's Next?* edited by David Hale and Lyric Hale, 139–150. New Haven: Yale University Press.

Estadão. 2010. "China compra terras no Brasil." *Estadão de S. Paulo*, August 3.

European Society of Cardiology. 2016. "Obesity Trebles in Brazil Schoolchildren over 30 years." Press release, September 23. http://www.escardio.org/The-ESC/Press-Office/Press-releases/obesity-trebles-in-brazil-schoolchildren-over-30-years.

Evans, Peter. 1996. "Introduction: Development Strategies Across the Public-Private Divide." *World Development* 24 (6): 1033–1037.

FAO (United Nations Food and Agriculture Organization). 2003. *World Agriculture: Towards 2015/2030—An FAO Perspective*. Rome: FAO.

FAO. 2014. *Growing Greener Cities in Latin America and the Caribbean.* Rome: FAO.

Farrelly, Kate. 2018. "Food for Thought: Sustainable Eating Becomes Property Selling Point Across Australia." Domain Real Estate website, October 15. http://www.domain.com.au/news/simple-vegie-patch-becoming-property-selling-point-773181/.

Feinberg, Richard E. 2016. *Open for Business: Building the New Cuban Economy*. Washington, DC: Brookings Institution Press.

Fernandes, Evandro Noro. 2008. "Reprodução de pequenos agricultores no espaço metropolizado paulistano: Uma análise da porção sul do município de São Paulo." Unpublished Masters thesis, Department of Geography, University of São Paulo.

Fernández, Avelino. 2006. *Lecturas en pro del cooperativismo: Ante las imprescindibles transformaciones económicas del socialismo cubano.* Cienfuegos: Universo Sur.

Ferreira Abrão, Rafael Almeida, and M. P. Amineh. 2024. "Brazilian Perception of the China-Led Belt and Road Initiative." *Journal of Contemporary China* 33 (150): 987–1005. https://doi.org/10.1080/10670564.2023.2299792.

Ferreira, Jamile. 2015. "Caracterização dos agricultores orgânicos das feiras da cidade de São Paulo." Unpublished Masters thesis, Department of Agrarian Sciences, Federal University of Paraná.

FGV (Fundación Getulio Vargas). 2023. Wealth Map: New Data Shows That Brasilia Has Highest Income in Brazil, February 14. https://portal.fgv.br/en/news/wealth-map-new-data-shows-brasilia-has-highest-income-brazil.

Fornet Piña, Fernando. 2015. *Fiesta gastronòmica de la mesa tradicional cubana.* Havana: Editorial Científico-Técnica.

Fox, Jonathan. 1996. "How Does Civil Society Thicken? The Political Construction of Social Capital in Rural Mexico." *World Development* 24 (6): 1089–1103.

França, Carlos Arthur. 2011. "Concentração De Áreas Verdes Cria 'Apartheid Vegetal.'" Folha de São Paulo, July 24. http://classificados.folha.uol.com.br/imoveis/948671-concentracao-de-areas-verdes-cria-apartheid-vegetal.shtml.

Freidberg, Susanne. 2004. *French Beans and Food Scares: Culture and Commerce in an Anxious Age.* Oxford and New York: Oxford University Press.

Friedmann, Hariett, and Philip McMichael. 1989. "Agriculture and the State System: The Rise and Fall of National Agricultures, 1870 to the Present." *Sociologia Ruralis* 29 (2): 93–117.

Fukuyama, Francis. 1992. *The End of History and the Last Man.* New York: Free Press.

Fukuyama, Francis. 1995. *Trust: The Social Virtues and the Creation of Prosperity.* New York: Free Press.

Funes, Fernando, Luis Garcia, Martin Bourque, Nilda Perez, and Peter Rosset. 2002. *Sustainable Agriculture and Resistance: Transforming Food Production in Cuba.* Oakland, CA: Food First Books.

Funes Monzote, Fernando R. 2008. *Farming Like We're Here to Stay: The Mixed Farming Alternative for Cuba.* Wageningen: University of Wageningen.

Funes Monzote, Fernando R. 2019. *Nuestro viaje a la Luna: La idea de la transformación de la naturaleza en Cuba durante la Guerra Fría.* Havana: Fondo Editorial Casa de las Américas.

Galvin, Shayla. 2021. *Becoming Organic.* New Haven: Yale University Press.

Gamboa Costa, Abel, Roberto González Sousa, and Angelina Herrera Sorzano. 2013. "Soberanía y seguridad alimentaria en Cuba: Políticas públicas necesarias para reducir la dependencia alimentaria." *Agrisost* 19 (3): 1–14.

García Triana, Mauro, Pedro Eng Herrera, and Gregor Benton. 2009. *The Chinese in Cuba: 1847–Now.* Lanham, MD: Lexington.

Gardiner, Ed. 2017. "Backing for Community Garden in Pentridge Prison Development." *The Herald Sun*, January 10. http://www.heraldsun.com.au/leader/north-west/backing-for-community-garden-in-pentridge-prison-development/news-story/13f54bc0d1a06022f34289286662527b.

Garth, Hannah. 2013. Obesity in Cuba: Memories of the Special Period and Approaches to Weight Loss Today. In *Reconstructing Obesity: The Meaning of Measures and the Measure of Meanings*, edited by Megan B. McCullough and Jessica A. Hardin, 89–106. New York: Berghahn Books.

Garth, Hannah. 2020. *Food in Cuba: The Pursuit of a Decent Meal*. Stanford: Stanford University Press.

Gaynor, Andrea. 2006. *Harvest of the Suburbs: An Environmental History of Growing Food in Australian Cities*. Perth: University of Western Australia Press.

Gaynor, Andrea. 2018. "Learning from Our Productive Past." In *Reclaiming the Urban Commons: The Past, Present and Future of Food Growing in Australian Towns and Cities*, edited by Nick Rose and Andrea Gaynor, 165–172. Perth: University of West Australia Press.

Geertz, Clifford. C. 1973. *The Interpretation of Cultures: Selected Essays*. New York: Basic.

Giacchè, Giulia, and Wania Resende Silva. 2014. "Urban Agriculture in Sao Paulo (Brazil): Actors, Space and Governance Models." Paper presented at 6th AESOP Conference on Sustainable Food Planning.

Global Exchange. 2019. Reality Tours: Sustainability in Cuba. https://app.anyguide.com/i/realitytours/c/tours/the-sustainability-and-community-tour-of-cuba.

Global Times. 2021. "China, Cuba Sign Cooperation Plan to Promote BRI Construction." Editorial, December 26. https://www.globaltimes.cn/page/202112/1243371.shtml.

Global Victoria. 2024. Greater China. https://global.vic.gov.au/explore-our-global-markets/china.

Globo. 2016. "Hortas comunitárias que ficam perto do trânsito também sofrem os efeitos da poluição." June 16. https://globoplay.globo.com/v/5095470/.

Goffman, Erving. 1969. *The Presentation of Self in Everyday Life*. London: The Penguin Press.

Gondal, Neha. 2022. "Multiplexity as a Lens to Investigate the Cultural Meanings of Interpersonal Ties." *Social Networks* 68(2): 209–217.

González, Lenier. 2020. "Cuba 2020: ¿a las puertas de una crisis humanitaria?" The Inter-American Dialogue. *The Latin America Advisor.* Accessed 11-30-2020. https://www.thedialogue.org/blogs/2020/04/cuba-2020-a-las-puertas-de-una-crisis-humanitaria/.

Gorenstein, Ossir. 2008. "Monitoramento de resíduos de agrotóxicos realizado pela CEAGESP no período de 2006 a maio de 2007." *Informações Econômicas São Paulo* 38 (6): 20-28.

Gould, Kenneth A., and Tammy L. Lewis. 2017. *Green Gentrification: Urban Sustainability and the Struggle for Environmental Justice.* Abingdon and New York: Routledge.

Governo do Brasil. 2017. "Obesidade cresce 60% em dez anos no Brasil." April 17. http://www.brasil.gov.br/saude/2017/04/obesidade-cresce-60-em-dez-anos-no-brasil.

Governo do Brasil. 2023. Lista e íntegra dos atos assinados no Grande Palácio do Povo, por ocasião da visita do Presidente Luiz Inácio Lula da Silva à República Popular da China. Nota à imprensa no.131. https://www.gov.br/mre/pt-br/canais_atendimento/imprensa/notas-a-imprensa/lista-e-integra-dos-atos-assinados-no-grande-palacio-do-povo-por-ocasiao-da-visita-do-presidente-luiz-inacio-lula-da-silva-a-republica-popular-da-china.

Granovetter, Mark. 1973. "The Strength of Weak Ties." *American Journal of Sociology* 78 (6): 1360–1380.

Griffith, Lauren Miller. 2018. "Hungry, Hungry Gringos: Solving the Problems Posed by Feeding Tourists." *Culture, Agriculture, Food and Environment* 41 (1): 2433.

Gro Intelligence. 2017. "Brazilian Soybeans and the Amazon Rainforest." 6th September. https://gro-intelligence.com/insights/brazil-soybeans-amazon-rainforest.

Grupo SP. 2007. "Edifício residencial na Rua Simpatia." http://www.gruposp.arq.br/?p=19.

Guangming. 2023. "Interview with Wang Yiwei: Overcoming the Paradox of Globalization, the 'Belt and Road' Boosts Global Development and Prosperity." *Guangming Net—Academic Channel,* October 13. https://share.gmw.cn /www/xueshu/2023-10/13/content_36891637.htm.

Gunn, Gillian. 1995. *Cuba's NGOs: Government Puppets or Seeds of Civil Society?* Washington, DC: Center for Latin American Studies, Georgetown University.

Gürcan, Efe Can. 2014. "Cuban Agriculture and Food Sovereignty: Beyond Civil-Society-Centric and Globalist Paradigms." *Latin American Perspectives* 197 (41–4): 129–146.

Hall, Tony. 2010. *The Life and Death of the Australian Backyard.* Collingwood, Victoria: CSIRO Publishing.

Hansing, Katrin, and Bert Hoffmann. 2020. "When Racial Inequalities Return: Assessing the Restratification of Cuban Society 60 Years After Revolution." *Latin American Politics and Society* 62 (2): 29–52.

Hayano, David M. 1979. "Auto-Ethnography: Paradigms, Problems and Prospects." *Human Organization* 38 (1): 99–104.

Hearn, Adrian H. 2004a. "Afro-Cuban Religions and Social Welfare: Consequences of Commercial Development in Havana." *Human Organization* 63 (1): 78–87.

Hearn, Adrian H. 2004b. "Political Dimensions of International NGO Collaboration with Cuba." In *Cuba Today: Continuity and Change Since the 'Special Period'*, edited by Mauricio Font, 209–227. New York: CUNY Graduate Center.

Hearn, Adrian H., ed. 2008a. *Cultura, Tradición, y Comunidad: Perspectivas sobre la Participación y el Desarrollo en Cuba*. Havana: Imagen Contemporánea and UNESCO International Center for Sustainable Human Development.

Hearn, Adrian H. 2008b. *Cuba: Religion, Social Capital, and Development*, Durham, NC: Duke University Press.

Hearn, Adrian H. 2012. "What Can South America and Australia Learn from Each Other in Relation to China?" In *Setting the Agenda: Asia and Latin America in the 21st Century*, edited by Ariel C. Armony, 33–38. Miami: University of Miami.

Hearn, Adrian H. 2013. *China and Brazil: Searching for Sustainable Complementarity*. The European Commission—Europe China Research and Advice Network policy paper #72. London: ECRAN.

Hearn, Adrian H. 2016. *Diaspora and Trust: Cuba, Mexico, and the Rise of China*. Durham, NC: Duke University Press.

Hearn, Adrian H. 2019. "Chinese Investment in Australia: Sustainable Diversification?" In *China's Foreign Direct Investment in Latin America and the Caribbean: Conditions and Challenges*, edited by Enrique Dussel Peters, 87–102. Boulder: Lynne Rienner Publishers.

Hearn, Adrian H. 2023. "Bringing Community Together Through Urban Farming: Rio de Janeiro." *Pursuit*, July 28. https://pursuit.unimelb.edu.au/articles/bringing-community-together-through-urban-farming.

Hearn, Adrian H., and José Luis León Manríquez, eds. 2011. *China Engages Latin America: Tracing the Trajectory*. Boulder: Lynne Rienner Publishers.

Hearn, Adrian H., and Rafael Hernández. 2018. "China and Cuba." In *Cuba's Foreign Relations Under Raúl Castro*, edited by John Kirk and Henry Erisman, 255–267. Boulder: Roman and Littlefield.

Hearn, Adrian H., and Rafael Hernández. 2021. "Cuba-China Relations and the Construction of Socialism." In *Cuban International Relations at 60: Reflections on Global Connections*, edited by Mervyn Bain and Chris Walker, 39–156. Lanham: Lexington Books.

Hellyer, Marcus, and Andrew Nicholls. 2022. *'Impactful Projection': Long-Range Strike Options for Australia*. Canberra: Australian Strategic Policy Institute.

Hempel, Corinna, and Ulrich Hamm. 2016. "How Important is Local Food to Organic-Minded Consumers?" *Appetite* 96: 309–318.

Hermans, Theo, ed. 1985. *The Manipulation of Literature*. London: Routledge.

Hernández Cáceres, Alberto. 2020. "Avanza en Pinar del Río construcción de moderno secadero de arroz." *Radio Guamá*, July 16. https://www.radiosandino.icrt.cu/noticias/secadero-de-arroz-pina/.

Hernández, Rafael. 2003. *Looking at Cuba: Essays on Culture and Civil Society*. Gainesville: University Press of Florida.

Hillman, Jonathan E. 2018. "China's Belt and Road Is Full of Holes." *CSIS Briefs*, September. Washington, DC: CSIS.

Hobsbawm, Eric, and Terence Ranger. 1992. *The Invention of Tradition*. Cambridge, UK: Cambridge University Press.

Holmgren, David. 2018. *Retrosuburbia: A Downshifter's Guide to a Resilient Future*. Hepburn Springs, VIC: Melliodora.

Holt-Giménez, Eric. 2006. *Campesino a Campesino: Voices from Latin America's Farmer to Farmer Movement for Sustainable Agriculture*. Oakland, CA: Food First.

Holt-Giménez, Eric. 2009. "From Food Crisis to Food Sovereignty: The Challenge of Social Movements." *Monthly Review* 61 (3): 142-156.

Holt-Giménez, Eric. 2011. "Food Security, Food Justice, or Food Sovereignty? Crises, Food Movements, and Regime Change." In *Cultivating Food Justice: Race, Class, and Sustainability*, edited by Alison Hope Alkon and Julian Agyeman, 309–330. Cambridge, MA: MIT Press.

Holt-Giménez, Eric, and Annie Shattuck. 2011. "Food Crises, Food Regimes and Food Movements: Rumblings of Reform or Tides of Transformation?" *Journal of Peasant Studies* 38 (1): 109–144.

Holt-Giménez, Eric, Raj Patel, and Annie Shattuck. 2009. *Food Rebellions! Crisis and the Hunger for Justice*. Boston, MA: Pambazuka Press.

Hope Alkon, Alison, and Julian Agyeman. 2011. "Introduction: The Food Movement as Polyculture." In *Cultivating Food Justice: Race, Class, and Sustainability*, edited by Alison Hope Alkon and Julian Agyeman, 1–20. Cambridge, MA: MIT Press.

Horn, Sebastian, Bradley C. Parks, Carmen M. Reinhart, and Christoph Trebesch. 2023. "China as an International Lender of Last Resort." World Bank Policy Research Working Paper 10380. The World Bank. https://documents1.worldbank.org/curated/en/099450403272313885/pdf/IDU046bbbd8d06cc0045a708397004cbf4d2118e.pdf.

Howard, Ebenezer. 1902. *Garden Cities of To-Morrow*. London: Swan Sonnenschein and Co.

Huma. 2017. Futuro lançamento Huma Itaim. Video promotion (3:30–4:05). http://www.huma.net.br/empreendimento/huma-itaim/.

IAASTD (International Assessment of Agricultural Knowledge, Science and Technology for Development). 2009. *Agriculture at a Crossroads: Global Report*. Washington, DC: Island Press.

Ideazarvos. 2019. Nossa nova conquista em 2018. https://ideazarvos.com.br/pt/2018/11/22/ideazarvos-decorado-uma-nova-experiencia/.

IIPJM (Instituto Internacional de Periodismo José Martí). 2021. "#HerramientasParaPeriodistas: Soberanía alimentaria." October 14. https://periodismojosemarti.wordpress.com.

Jabbour, Elias. 2022. "Miguel Diaz-Canel's Strategic Visit Enhances China-Cuba Ties." *China Global Television Network*, November 29. https://news.cgtn.com/news/2022-11-29/Miguel-Diaz-Canel-s-strategic-visit-enhances-China-Cuba-ties-1flNmAxKGhq/index.html.

Jacobs, Andrew, and Matt Richtel. 2017. "How Big Business Got Brazil Hooked on Junk Food." *The New York Times*, September 16. http://www.nytimes.com/interactive/2017/09/16/health/brazil-obesity-nestle.html.

Jaffee, Daniel. 2007. *Brewing Justice: Fair Trade Coffee, Sustainability, and Survival*. Berkeley: University of California Press.

Jatar-Haussmann, Ana Julia. 1999. *The Cuban Way*. West Hartford: Kumarian Press.

Jiang, Shixue. 2011. "Ten Key Questions." In *China Engages Latin America: Tracing the Trajectory*, edited by Adrian H. Hearn and José Luis León Manríquez, 51–65. Boulder: Lynne Rienner Publishers.

Jordahl, Henrik. 2007. "Inequality and Trust." IFN Working Paper No.715, Research Institute of Industrial Economics.

Joseph, Sabrina, ed. 2019. *Commodity Frontiers and Global Capitalist Expansion: Social, Ecological and Political Implications from the Nineteenth Century to the Present Day*. London: Palgrave Macmillan.

Kawa, Nicholas C., Christopher McCarty, and Charles R. Clement. 2013. "Manioc Varietal Diversity, Social Networks, and Distribution Constraints in Rural Amazonia." *Current Anthropology* 54 (6): 764–770.

Koont, Sinan. 2009. "The Urban Agriculture of Havana." *Analytical Monthly Review* 60 (8): 44–63.

Labrador Machín, Odalys, Yamira Mirabal González, and Carlos Torres Paez. 2020. "Potential of Agro-Livestock Cooperatives in Terms of Food Sovereignty and Nutritional Education." *Cooperativismo y Desarrollo* 8 (3): 587–602.

Lahmeyer Lobo, Eulália Maria. 1978. *História do Rio de Janeiro*. Rio de Janeiro: Access Editora.

Lam, Willy. 2004. "China's Encroachment on America's Backyard." *China Brief* 4 (23): 1–3.

Latour, Bruno. 2015. *Reassembling the Social: An Introduction to Actor-Network Theory*. New York: Oxford University Press.

Lazarus, Leland, and Evan Ellis. 2021. "How China Helps the Cuban Regime Stay Afloat and Shut Down Protests." *The Diplomat*, August 3. https://thediplomat.com/2021/08/how-china-helps-the-cuban-regime-stay-afloat-and-shut-down-protests/.

Le Polain de Waroux, Yann, Matthias Baumann, Nestor Ignacio Gasparri, Gregorio Gavier-Pizarro, Javier Godar, Tobias Kuemmerle, Robert Müller, Fabricio Vázquez, José Norberto Volante, Patrick Meyfroidt. 2018. "Rents, Actors, and the Expansion of Commodity Frontiers in the Gran Chaco." *Annals of the American Association of Geographers* 108 (1): 204–225.

Leahy, Joe. 2017. "São Paulo mayor touted as possible strong presidential candidate." *The Financial Times*, May 16. https://www.ft.com/content/1a3fe66a-3929-11e7-821a-6027b8a20f23.

Leahy, Joe, James Kynge, and Benjamin Parkin. 2023. "Ten years of China's Belt and Road: What Has $1tn Achieved?" *Financial Times*, October 22. https://www.ft.com/content/83501dd5-fe6d-4169-9d83-28a8cf46e681.

Ledeneva, Alena. 1998. *Russia's Economy of Favors: Blat, Networking and Informal Exchange*. Cambridge: Cambridge University Press.

Leite, Rogerio. 2015. "Cities and Gentrification in Contemporary Brazil." *Current Urban Studies* 3 (3): 175–186.

Leitgeb, Friedrich, Sarah Schneider, and Christian R. Vogl. 2016. "Increasing Food Sovereignty with Urban Agriculture in Cuba." *Agriculture and Human Values* 33 (2): 415–426.

Lévi-Strauss, Claude. 1962. *La Pensée sauvage*. Paris: Librarie Plon. https://ia801607.us.archive.org/22/items/lapenseesauvage00levi/lapenseesauvage00levi.pdf.

Li, Lilian. 2007. *Fighting famine in North China*. Stanford: Stanford University Press.

Lin, Nan. 2001. *Social Capital: A Theory of Social Structure and Action*. Cambridge: Cambridge University Press.

Lin, Scott Y. 2022. "Restoring the State Back to Food Regime Theory: China's Agribusiness and the Global Soybean Commodity Chain." *Journal of Contemporary Asia* 53 (2): 288–310.

Lindsay, Nicole. 2016. "Stockland Ramps Up Green Stars." *The Sydney Morning Herald*, August 31. http://www.smh.com.au/business/property/stockland-ramps-up-green-stars-20160830-gr4abq.html.

Little Projects. 2015. "Urban Renewal: Tip Top Brunswick East." http://www.littleprojects.com.au/2015/03/30/urban-renewal-tip-top-brunswick-east/.

Liu, Dong. 2017. "Tipping the Scale: Beijing Leads in Obesity Rate." *The China Daily*, June 28. http://www.chinadaily.com.cn/china/2017-06/28/content_29921273.htm.

Lloyd, Peter. 2018. "Why We Should Worry About Victoria's China Memorandum of Understanding." *The Conversation*, November 21.

Lo, Ming-Cheng M., and Eileen M. Otis. 2003. "Guanxi Civility: Processes, Potentials, and Contingencies." *Politics and Society* 31 (1): 131–162.

Longbottom, Jessica, and Ben Knight. 2018. "Melbourne's Population Explosion Threatens to Create a 'Bangkok Situation.'" *ABC News*, October 16. http://www.abc.net.au/news/2018-10-15/melbourne-will-be-australias-biggest-city-which-party-has-policy/10358988.

López-Doval, Julio C., Cassiana C. Montagner, Anjaína Fernandes de Alburquerque, Viviane Moschini-Carlos, Gisela Umbuzeiro, and Marcelo Pompêo. 2017. "Nutrients, Emerging Pollutants and Pesticides in a Tropical Urban Reservoir: Spatial Distributions and Risk Assessment." *Science of The Total Environment* 575: 1307–1324.

Losos, Elizabeth, Alexander Pfaff, and Lydia Olander. 2019. "The Deforestation Risks of China's Belt and Road Initiative." *Brookings Future Development*, January 28. https://www.brookings.edu/blog/future-development.

Lowry, Nelson. 1950. *Rural Cuba*. Minneapolis: University of Minnesota Press.

Lowy Institute for International Policy. 2023. Lowy Institute Poll Themes: China. https://poll.lowyinstitute.org/themes/china.

Lucas, Paola. 2017. "Marcelo Crivella anuncia expansão de hortas comunitárias." *O Dia*, January 7. http://odia.ig.com.br/rio-de-janeiro/2017-01-07/marcelo-crivella-anuncia-expansao-de-hortas-comunitarias.html.

Lyon, Phil, and Anne Colquhoun. 1999. "Selectively Living in the Past: Nostalgia and Lifestyle." *Journal of Consumer Studies & Home Economics* 23 (3): 191–196.

Lyon, Sarah. 2011. *Coffee and Community: Maya Farmers and Fair-Trade Markets*. Boulder, CO: University Press of Colorado.

Lyson, Thomas, G.W. Stevenson, and Rick Welsh, eds. 2008. *Food and the Mid-Level Farm: Renewing an Agriculture of the Middle*. Cambridge, MA: MIT Press.

Lyu, Lily, Jin Yang, and Wang Xiumin. 2020. "New Money: CSA Farms Offer Inspiration to China's Agriculture." CGTN, May 22. https://news.cgtn.com/news/2020-05-22/New-Money-CSA-farms-offer-inspiration-to-China-s-agriculture-QHBXoPCjIc/index.html.

Machín Sosa, Braulio, Adilén María Roque Jaime, Dana Rocío Ávila Lozano, and Peter Michael Rosset. 2010. *Revolución agroecológica*. Havana: ANAP.

Maciel, Rodrigo, and Dani K. Nedal. 2011. "China and Brazil: Two Trajectories of a 'Strategic Partnership.'" In *China engages Latin America: Tracing the Trajectory*, edited by Adrian H. Hearn and José Luis León-Manríquez, 235–255. Boulder: Lynne Rienner.

MacRae, Graeme. 2016. "Food Sovereignty and the Anthropology of Food: Ethnographic Approaches to Policy and Practice." *Anthropological Forum* 26 (3): 227–232.

Madeira, Nuno R., Francisco J. B. Reifschneider, and Leonardo de B. Giordano. 2008. "Contribuição portuguesa à produção e ao consumo de hortaliças no Brasil: Uma revisão histórica." *Horticultura Brasileira* 26: 428–432.

Magalhaes, Luciana, and Samantha Pearson. 2017. "São Paulo Mayor João Doria Puts City's Assets Up for Sale." *Wall Street Journal*, January 25. https://www.wsj.com/articles/sao-paulo-mayor-joao-doria-puts-citys-assets-up-for-sale-1485366255.

Malinowski, Bronislaw. 1922. *Argonauts of the Western Pacific*. London: Routledge.

Mao, Xianglin, Carlos Alzugaray Treto, Liu Weiguang, and Adrian H. Hearn. 2011. "China and Cuba: Past, Present, and Future." In *China Engages Latin America: Tracing the Trajectory*, edited by Adrian H. Hearn and José Luis León Manríquez, 187–201. Boulder: Lynne Rienner Publishers.

Martí, José. 1963. *Obras Completas*. La Habana: Editora Nacional de Cuba.

MCC (Moreland City Council). 2017. *Moreland Food System Strategy 2017–2020*. Moreland: MCC.

MCC. 2020. *Moreland Food System Strategy Extension 2020–2024*. Moreland: MCC.

McCarthy, Breda, and Hong Bo Liu. 2017. "Food Waste and the 'Green' Consumer." *Australasian Marketing Journal* 25 (2): 126–132.

McClintock, Nathan. 2014. "Radical, Reformist, and Garden-Variety Neoliberal: Coming to Terms with Urban Agriculture's Contradictions." *Local Environment* 19 (2): 147–171.

McClintock, Nathan. 2017. "Cultivating (a) Sustainability Capital: Urban Agriculture, Ecogentrification, and the Uneven Valorization of Social Reproduction." *Annals of the American Association of Geographers* 108 (2): 579–590.

McMichael, Philip. 2014. "Historicizing Food Sovereignty." *Journal of Peasant Studies* 41 (6): 933–957.

MDAS (Ministério do Desenvolvimento e Assistência Social, Família e Combate à Fome). 2023. "MDS e MDA traçam estratégia com hortas urbanas para tirar o Brasil do mapa da fome." April 20. https://www.gov.br/mds/pt-br/noticias-e-conteudos/desenvolvimento-social/noticias-desenvolvimento-social/mds-e-mda-tracam-estrategia-com-hortas-urbanas-para-tirar-o-brasil-do-mapa-da-fome.

Millar, Royce, Chris Vedelago, and Bianca Hall. 2020. "What happens to Melbourne when it's starved of the growth that fed its lifestyle?" *The Age*, October 16. https://www.theage.com.au/national/victoria/what-happens-to-melbourne-when-it-s-starved-of-the-growth-that-fed-its-lifestyle-20201015-p565m6.html.

Ministry of Foreign Affairs of Cuba. 1960. "Memorandum of the Conversation between Premier Zhou Enlai and Cuban Revolutionary Government Economic Delegation, November 18, 1960." Available from the Wilson Center History and Public Policy Program Digital Archive, PRC fma 204-00098-02,1–16. Translated by Zhang Qian. http://digitalarchive.wilsoncenter.org/document/115156.

Ministry of Health of Brazil. 2016. Vigilância de fatores de risco e proteção para doenças crônicas por inquérito telefônico. Brasilia: Ministério da Saúde do Brasil.

Mintz, Sidney. 1985. *Sweetness and Power: The Place of Sugar in Modern History*. New York: Viking.

Miskimmon, Alister, Ben O'Loughlin, and Jinghan Zeng. 2021. *One Belt, One Road, One Story?* London: Palgrave Macmillan.

Mollison, Bill, and David Holmgren. 1978. *Permaculture One*. Australia and New Zealand: Corgi Publishing.

Moon, Krystyn R., Jennifer Rhode Ward, José Vazquez Rodriguez, and Jorge Foyo. 2021. "La Picadora: A Case Study in Cuban Agrotourism." *International Journal of Cuban Studies* 13 (1): 8–42.

Moon, Krystyn R., Jennifer Rhode Ward, José Vazquez Rodriguez, and Jorge Foyo. 2022. "Food Access, Identity, and Taste in Two Rural Cuban Communities." *Gastronomica: The Journal for Food Studies* 22 (1): 66–78.

Morales, Emilio. 2019. "Remesas, ¿una ruta de inversión para los cubanos?" The Havana Consulting Group, September 27. Accessed 11-30-2020. http://www.thehavanaconsultinggroup.com/es-es/Articles/Article/69.

Moreira, João Carlos. 2017. "Inovação em marketing transporta a Mata Atlântica ao Brás." *Estadão*, August 31. https://economia.estadao.com.br/noticias/seu-imovel,inovacao-em-marketing-transporta-a-mata-atlantica-ao-bras,70001956831.

Morellato, L. Patricia, and Célio F. B. Haddad. 2000. "Introduction: The Brazilian Atlantic Forest." *Biotropica* 32 (4b): 786–792.

Morgan, Faith. (Director). 2006. *The Power of Community: How Cuba Survived Peak Oil*. Yellow Springs: AlchemyHouse Productions Inc. and The Community Solution.

Morgan, Henry L. 1871. *Systems of Consanguinity and Affinity of the Human Family.* Smithsonian Contributions to Knowledge, vol. 17. Washington: Smithsonian Institution.

MST (Movimento dos Trabalhadores Rurais Sem Terra). 2021. "Você sabe o que é Soberania Alimentar?" MST Podcast series, October 13. https://mst.org.br/2021/10/13/voce-sabe-o-que-e-soberania-alimentar-2/.

Muniz, Carolina. 2017. "Empresa permite adotar galinha e cuidar de horta a distância." *Folha de São Paulo,* August 7.

Murphy, Maureen, Rachel Carey, and Leila Alexandra. 2022. *The Resilience of Melbourne's Food System to Climate and Pandemic Shocks.* University of Melbourne: Victorian Eco-Innovation Lab.

Myers, Margaret, and Carol Wise, eds. 2017. *The Political Economy of China-Latin America Relations in the New Millennium: Brave New World.* New York: Routledge.

Myers, Margaret, Lin Hua, Jorge Heine, and Thomas Rideg. 2023. "What Is the State of Chinese Investment in Latin America?" *Latin America Advisor,* September 8. https://www.thedialogue.org/analysis/what-is-the-state-of-chinese-investment-in-latin-america/.

Nagib, Gustavo. 2016. "Agricultura urbana como ativismo na cidade de São Paulo: o caso da Horta das Corujas." Unpublished Masters degree, University of São Paulo.

Nagib, Gustavo. 2018. *Agricultura urbana como ativismo na cidade de São Paulo.* São Paulo: Annablume.

Nagib, Gustavo, and Angélica Campos Nakamura. 2020. "Urban Agriculture in the City of São Paulo: New Spatial Transformations and Ongoing Challenges to Guarantee the Production and Consumption of Healthy Food." *Global Food Security* 26: 1–7.

NCUSCR (National Committee on US-China Relations) 2023. China's Growing Influence in Latin America: A Close Look at China-Cuba Relations. August 23. https://www.ncuscr.org/video/china-cuba-relations/.

NDRC (National Development and Reform Commission). 2013. *Report on the Implementation of the 2012 Plan for National Economic and Social Development and on the 2013 Draft Plan for National Economic and Social Development.* Beijing: NDRC.

NDRC. 2023. *Report on the Implementation of the 2022 Plan for National Economic and Social Development and on the 2023 Draft Plan for National Economic and Social Development.* Beijing: NDRC.

Neill, Casey. 2016. "How Sweet It Is!" *Dandenong Journal,* May 17. http://dandenong.starcommunity.com.au/journal/2016-05-17/how-sweet-it-is-2/.

Nelson, Erin, Steffanie Scott, Judie Cukier, and Ángel Leyva Galán. 2009. Institutionalizing Agroecology: Successes and Challenges in Cuba. *Agriculture and Human Values* 26 (3): 233–243.

Nonini, Donald M. 2014. "Food and Farming." In *A Companion to Urban Anthropology*, edited by Donald M. Nonini, 394–413. Chichester, UK: John Wiley & Sons.

Nugent, Jeffrey B. 1993. "Between State, Market and Households: A Neoinstitutional Analysis of Local Organizations and Institutions." *World Development* 21 (4): 623–32.

Nuttavuthisit, Krittinee, and John Thøgersen. 2017. "The Importance of Consumer Trust for the Emergence of a Market for Green Products: The Case of Organic Food." *Journal of Business Ethics* 140 (2): 323–337.

OEC (Observatory of Economic Complexity). 2021. Brazil. https://oec.world/en/profile/country/bra.

OECD (Organization for Economic Cooperation and Development). 2023. Meat Consumption. https://data.oecd.org/agroutput/meat-consumption.htm.

Ofstehage, Andrew. 2011. "Nusta Juira's Gift of Quinoa: Peasants, Trademarks, and Intermediaries in the Transformation of a Bolivian Commodity Economy." *Anthropology of Work Review* 32 (2): 103–114.

Oliveira, Francisco. 2006. "Lula in the Labyrinth." *New Left Review* 42: 5–22.

Oliveira, Gustavo. 2021. "China's Engagement with Brazil since 2000: Key Actors, Strategies, and Conflicts." Presentation at the Hong Kong University of Science and Technology, March 9.

Oliveira Jardim, Andreia Nunes, and Eloisa Dutra Caldas. 2012. "Brazilian Monitoring Programs for Pesticide Residues in Food—Results from 2001 to 2010." *Food Control* 25 (2): 607–616.

Oppenheimer, Andres. 1992. *Castro's Final Hour: An Eyewitness Account of the Disintegration of Castro's Cuba*. New York: Touchstone.

O'Reilly, Érika de Mattos. 2014. "Agricultura Urbana—Um Estudo de Caso do Projeto Hortas Cariocas em Manguinhos, Rio de Janeiro." Academic thesis, Universidad Federal do Rio de Janeiro.

Organics News Brazil. 2017. "Mercado de orgânicos cresce o dobro no Brasil," June 19. http://www.organicsnewsbrasil.com.br/negocio/agricultura/numero-de-produtores-organicos-cresce-cerca-de-17-no-estado-de-sao-paulo/.

Organicsnet. 2016. "Mercado de orgânicos cresce o dobro no Brasil." January 11. http://www.organicsnet.com.br/2016/01/mercado-de-organicos-cresce-o-dobro-no-brasil/.

Ortiz, Fernando. 1995. *Cuban Counterpoint: Tobacco and Sugar*. Durham: Duke University Press. Originally published in 1940.

Ortner, Sherry. B. 1991. "Patterns of History: Cultural Schemas in the Founding of Sherpa Religious Institutions." In *Culture Through Time: Anthropological Approaches*, edited by E. Ohnuki-Tierney, 57–93. Stanford: Stanford University Press.

Osterhoudt, Sarah, Shaila S. Galvin, Dana J. Graef, Alder K. Saxena, Michael R. Dove. 2020. "Chains of Meaning: Crops, Commodities, and the 'in-between' Spaces of Trade." *World Development* 350: 1–10.

Otero, Gerardo. 2018. *The Neoliberal Diet: Healthy Profits, Unhealthy People*. Austin: University of Texas Press.

Ozbun, T. 2023. "Area Planted with Soybean in Brazil from Crop Year 2010/11 to 2022/23." *Statista*, June 21. https://www.statista.com/statistics/741384/soybean-production-volume-brazil/.

Padrón Hernández, Maria. 2012. *Beans and Roses: Everyday Economies and Morality in Contemporary Havana, Cuba*. Gothenburg, SE: University of Gothenburg.

Pan, Tao. 2004. "Timeless Theme of International Relations." *Beijing Review* 47 (23).

Parham, Susan. 2005. "Designing the Gastronomic Quarter." *Architectural Design* 75 (3): 86–95.

Parham, Susan. 2015. *Food and Urbanism: The Convivial City and a Sustainable Future*. London: Bloomsbury Academic.

Parham, Susan. 2016. "Food and Garden Cities in Principle and Practice." The International Garden Cities Institute, Think Pieces, July. http://www.gardencitiesinstitute.com/think-piece/food-and-garden-cities-in-principle-and-practice.

Parliament of Australia. 1907. *Ex parte H.V. McKay* (Harvester Case), (1907) 2 CAR 1. Available at: https://www.fwc.gov.au/waltzing-matilda-and-the-sunshine-harvester-factory/historical-material/harvester-case.

Parsons, Levi, and Kylie Stevens. 2020. "Revealed: The Alarming Level of Chinese Ownership of Australia—From Cattle Stations, to Ports and Even our Water." *Daily Mail Australia*, December 3. https://www.dailymail.co.uk/news/article-9011631/The-shocking-level-Chinese-ownership-Australia-cattle-stations-ports-water.html.

Patel, Raj. 2009. "Food Sovereignty." *Journal of Peasant Studies* 36 (3): 663–706.

Payne, Marise. 2020. "Australia and the World in the Time of COVID-19." Speech to the National Security College, the Australian National University, June 16. https://www.foreignminister.gov.au/minister/marise-payne/speech/australia-and-world-time-covid-19.

Payne, Marise. 2021. "Decisions under Australia's Foreign Arrangements Scheme." Media release, April 21. https://www.foreignminister.gov.au/minister/marise-payne/media-release/decisions-under-australias-foreign-arrangements-scheme.

Pearsall, Hamil, and Isabelle Anguelovski. 2016. "Contesting and Resisting Environmental Gentrification: Responses to New Paradoxes and Challenges for Urban Environmental Justice." *Sociological Research Online* 21 (3): 1–7.

People's Daily. 1996. "Chairman Qiao Shi Met with Chairman Alarcon." Editorial, April 10.

People's Daily Video. 2020. The First "Belt and Road" Green Development Summit (English) 首届"一带一路"绿色发展大会 （英文下). http://v.people.cn/n1/2020/0926/c43911-31876013.html.

People's Republic of China Ministry of Foreign Affairs. 1987. A General Survey of Chinese Diplomacy. Beijing: World Affairs.

Pereira, Pedro Arraes, Geraldo B. Martha, Carlos A. Santana, and Eliseu Alves. 2012. "The Development of Brazilian Agriculture: Future Technological Challenges and Opportunities." *Agriculture & Food Security* 1 (4). https://doi.org/10.1186/2048-7010-1-4.

Pérez Sanchez, Amaury. 2020. "Cuba's First Biomass-Fired Power Plant Inaugurated." *Power Magazine*, July 1. https://www.powermag.com/cubas-first-biomass-fired-power-plant-inaugurated/.

Perfecto, Ivette, John Vandermeer, and Angus Wright. 2009. *Nature's Matrix: Linking Agriculture, Conservation and Food Sovereignty*. London: Earthscan.

Pertierra, Anna Cristina. 2012. "The More Things Change, the More They Stay the Same: Rice and Beans in Modern Cuba." In *Rice and Beans: A Unique Dish in a Hundred Places*, edited by Richard Wilk and Livia Barbosa, 35–60. London: Berg.

Phillips, Catherine, Olivia Dun, Ikerne Aguirre-Bielschowsky, Adrian Hearn, and Chris Williams. 2021. *Opportunities to Advance Food Gardening in Melbourne: Findings from the 2020 National Pandemic Gardening Survey*. Melbourne: Sustain.

Phillips, Dom. 2019. "Hundreds of New Pesticides Approved in Brazil under Bolsonaro." *The Guardian*, June 12, 2019.

Pollitt, Brian. 2004. "The Rise and Fall of the Cuban Sugar Economy." *Journal of Latin American Studies* 36 (2): 319–348.

PopulationStat. 2023a. Beijing, China Population. https://populationstat.com/china/beijing.

PopulationStat. 2023b. Rio de Janeiro, Brazil Population. https://populationstat.com/brazil/rio-de-janeiro.

Portes, Alejandro. 1998. "Social Capital: Its Origins and Applications in Modern Sociology." *Annual Review of Sociology* 24: 1–24.

Portes, Alejandro, and Ariel C. Armony. 2018. *The Global Edge: Miami in the Twenty-First Century.* Berkeley: University of California Press.

Poulston, Jill, and Albert Yau Kwong Yiu. 2011. "Profit or Principles: Why Do Restaurants Serve Organic Food?" *International Journal of Hospitality Management* 30 (1): 184–191.

Powell, Dawn. 2011. "The Dragon's Appetite for Soy Stokes Brazilian Protectionism." *The Financial Times*, October 10.

Powell, Kathy. 2008. "Neoliberalism, the Special Period and Solidarity in Cuba." *Critique of Anthropology* 28 (2): 177–197.

PRC State Council. 2023. *The Belt and Road Initiative: A Key Pillar of the Global Community of Shared Future.* Beijing: State Council Information Office of the People's Republic of China.

Prebisch, Raúl. 1950. *The Economic Development of Latin America and Its Principal Problems.* New York: United Nations.

Prefeitura de São Paulo. 2014. *City of São Paulo Strategic Master Plan.* São Paulo: Department of Urban Development.

Premat, Adriana. 2009. "State Power, Private Plots and the Greening of Havana's Urban Agriculture Movement." *City and Society* 21 (1): 28–57.

Premat, Adriana. 2012. *Sowing Change: The Making of Havana's Urban Agriculture.* Nashville, TN: Vanderbilt University Press.

Puntenney, Pamela J. 2009. "Preface." In *Global Ecosystems: Creating Options Through Anthropological Perspectives*, edited by Pamela J. Puntenney, 2–3. Washington, DC: The American Anthropological Association.

Putnam, Robert. 2000. *Bowling Alone: The Collapse and Revival of American Community.* New York: Simon and Schuster.

Putnam, Robert. 2007. "E Pluribus Unum: Diversity and Community in the Twenty-First Century." *Scandinavian Political Studies* 30 (2): 137–74.

Pym, Anthony. 2012. *On Translator Ethics: Principles for Cross-cultural Communication.* Amsterdam and Philadelphia: John Benjamins.

Qiao, Guanghua, Ting Guo, and Kurt Klein. 2012. "Melamine and Other Food Safety and Health Scares in China: Comparing Households with and Without Young Children." *Food Control* 26 (2): 378–386.

Radcliffe-Brown, Alfred. 1922. *The Andaman Islanders.* Cambridge: Cambridge University Press.

Reardon, Juan, and Reinaldo Pérez. 2010. "Agroecology and the Development of Indicators of Food Sovereignty in Cuban Food Systems." *Journal of Sustainable Agriculture* 34 (8): 907–922.

Rekow, Lea. 2015. "Fighting Insecurity: Experiments in Urban Agriculture in the Favelas of Rio de Janeiro." *Field Actions Science Reports* 8: 1–8.

Rekow, Lea. 2016. "On Unstable Ground: Issues Involved in Greening Space in the Rocinha Favela of Rio De Janeiro." *Journal of Human Security* 12 (1): 52–73.

República de Cuba. 2011. *Lineamientos de la política económica y social del partido y la revolución*. Havana: República de Cuba.

República de Cuba. 2021. *Proyecto de ley soberanía alimentaria y seguridad alimentaria y nutricional*. Havana: Asamblea Nacional del Poder Popular.

Ritter, Archibald R. M., and Ted A. Henken. 2015. *Entrepreneurial Cuba: The Changing Policy Landscape*. Boulder, CO: Lynne Rienner.

Rocha, Cassiano Brito, Ryan Nehring, and Sandro Dutra E. Silva. 2023. "Soy Without Borders: The Transnational Dynamics of Commodity Frontiers in South America (1971–2019)." *Global Environment* 15 (3). https://doi.org/10.3197/ge.2022.15030.

Rodrigues, Bernardo Salgado, and Elias Jabbour. 2023. "A iniciativa cinturão e rota e as implicações geoeconômicas para o Brasil sob a perspectiva da nova economia do projetamento e da geoeconomia híbrida." In *A China e a Iniciativa Cinturão e Rota: Percepçoes do Brasil*, edited by Evandro Menezes de Carvalho, Daniel Veras, and Pedro Steenhagen, 93–141. Rio de Janeiro: Getúlio Vargas Foundation.

Romero, Carlos A. 2023. "Cuba, China, Rusia y Venezuela: Los Límites de una Alianza Imperfecta." In *El Futuro de la Cuba Postrevolucionaria*, edited by Susanne Gratius and Matías Mongan, 292–310. Madrid: Tecnos.

Rosaldo, Renato. 1989. *Culture and Truth: The Remaking of Social Analysis*. Boston: Beacon Press.

Rose, Nick. 2013. "Optimism of the Will: Food Sovereignty as Transformative Counter-Hegemony in the 21st century." PhD diss., Royal Melbourne Institute of Technology (RMIT).

Rose, Nick, and Adrian H. Hearn. 2017. Food Systems and the Role of Local Government. Melbourne: SUSTAIN. http://www.circlesoffood.org/2016/04/18/food-governance-taskforce/.

Rose, Nick, and Andrea Gaynor, eds. 2018. *Reclaiming the Urban Commons: The Past, Present and Future of Food Growing in Australian Towns and Cities*. Perth: University of Western Australia Press.

Rosendahl, Mona. 1997. *Inside the Revolution: Everyday life in Socialist Cuba*. Ithaca, NY: Cornell University Press.

Rosito, Tatiana. 2020. "Bases para uma estratégia de longo prazo do Brasil para a China." Conselho Empresarial Brasil China, November 26. https://www.cebc.org.br/2020/11/26/bases-para-uma-estrategia-de-longo-prazo-do-brasil-para-a-china/.

Rosito, Tatiana, and Vinicius Mariano de Carvalho. 2022. *Brazil–China Trade Relations: In search of a strategy*. China in the World Policy Series. London: Kings College Lau China Institute.

Rosset, Peter M., and Medea Benjamin. 1994. *The Greening of the Revolution: Cuba's experiment with organic agriculture*. Melbourne: Ocean Press.

Rubino, Silvana. 2005. "A Curious Blend? City Revitalization, Gentrification and Commodification in Brazil." In *Gentrification in a global context: the new urban colonialism*, edited by Rowland Atkinson and Gary Bridge, 225–40. London: Routledge.

SABESP. 2017. Tratamento de Água—Região Metropolitana de São Paulo. http://site.sabesp.com.br/site/interna/Default.aspx?secaoId=36.

Sahlins, Marshall. 1981. *Historical Metaphors and Mythic Realities*. Ann Arbor: University of Michigan Press.

Sahlins, Marshall. 1999. "Two or Three Things That I Know about Culture." *Journal of the Royal Anthropological Institute* 5 (3): 399–422.

Salzman, Philip Carl. 2002. "On Reflexivity." *American Anthropologist* 104 (3): 805–813.

Samora, Roberto, and Ana Mano. 2020. "Brazil Deems Truck Stops 'Essential' to Keep Grains Flowing Despite Coronavirus." *Reuters*, March 28. https://www.reuters.com/article/health-coronavirus-brazil-logistics-idINL1N2BK0P7.

Sanjek, Roger. 1974. "What is Network Analysis and What Is It Good For?" *Reviews in Anthropology* 1 (4): 588–597.

Santandreu, Alain, and Ivana Cristina Lovo. 2007. *Panorama da agricultura urbana e periurbana no Brasil e diretrizes políticas para sua promoção*. Belo Horizonte: Rede de Intercâmbio de Tecnologias Alternativas.

Scheper-Hughes, Nancy. 1992. *Death Without Weeping: The Violence of Everyday Life in Brazil*. Berkeley: University of California Press.

Schnepf, Randall, Erik Dohlman, and Christine Bolling. 2001. "Agriculture in Brazil and Argentina: Developments and Prospects for Major Field Crops." USDA Economic Research Service, *Agriculture and Trade Report WRS-01-3*. Washington, DC: USDA. https://www.ers.usda.gov/publications/pub-details?pubid=40353.

Scholte, Bob. 1999. "Toward a Reflexive and Critical Anthropology." In *Reinventing Anthropology*, edited by Dell Hymes, 430-58. Ann Arbor: University of Michigan Press.

Schultz, Rainer. 2012. "Food Sovereignty and Cooperatives in Cuba's Socialism." *Socialism and Democracy* 26 (3): 117–138.

Schutte, Giorgio Romano. 2020. *Oásis para o capital—Solo fértil para a corrida de ouro: a dinâmica dos investimentos chineses no Brasil*. São Paulo: Editora Appris.

Schweizer, Thomas. 1997. "Embeddedness of Ethnographic Cases: A Social Network Perspective." *Current Anthropology* 38 (5): 739–760.

Siebert, Anne. 2023. *Food sovereignty and urban agriculture: concepts, politics, and practice in South Africa*. New York: Routledge.

Seto, Guilherme. 2017. "Câmara de SP aprova privatização de Interlagos em primeira votação." *Folha de São Paulo*, November 8. http://www1.folha.uol.com.br/cotidiano/2017/11/1933800-camara-de-sp-aprova-privatizacao-de-interlagos-em-primeira-votacao.shtml.

Seto, Guilherme. 2018. "Doria renuncia ao cargo, e Bruno Covas é o novo prefeito de São Paulo." *Folha de São Paulo*, April 6. https://www1.folha.uol.com.br/cotidiano/2018/04/doria-renuncia-ao-cargo-e-bruno-covas-e-o-novo-prefeito-de-sao-paulo.shtml.

Shambaugh, David. 2009. "The China awaiting President Obama." *Brookings East Asia Commentary*, November 10. http://www.brookings.edu/opinions/2009/11_china_shambaugh.aspx.

Shi, Yan. 2015. "BBC Dialogues: China's Organic Farming Pioneer, Shi Yan." *Fresh Dialogues: Interviews with Allison Van Diggelen*, July 14. http://www.freshdialogues.com/2015/07/14/bbc-dialogues-chinas-organic-farming-pioneer-shi-yan/.

Shim, Ji Young. 2018. "Problematic Autoethnographic Research: Researcher's Failure in Positioning." *The Qualitative Report* 23 (1): 1–11.

Shoebridge, Michael. 2020. "Victoria's Belt and Road Initiative Deal Undermines Cohesive National China Policy." *The Strategist*, Australian Strategic Policy Institute, May 22. https://www.aspistrategist.org.au/victorias-belt-and-road-initiative-deal-undermines-cohesive-national-china-policy/.

Singer, Paul. 1977. *Desenvolvimento econômico e evolução urbana*. São Paulo: Editora Nacional.

Sintusingha, Sidh, Hao Wu, Wenqi Lin, Sun Sheng Han, and Qin Bo. 2021. *International Perspectives on the Belt and Road Initiative*. London: Routledge.

Skocpol, Theda. 1996. "Unraveling from Above." *American Prospect* 7 (25): 20–25.

Smart, Alan, and Jinn Yuh Hsu. 2004. The Chinese Diaspora, Foreign Investment and Economic Development in China. *Review of International Affairs* 3 (4): 544–566.

Smart, Alan, and Josephine Smart. 2016. "China, Food Security, and the Trans-Pacific Partnership." In *The Changing Currents of Trans-Pacific Integration*, edited by Adrian H. Hearn and Margaret Myers, 99–113. Boulder: Lynne Rienner Publishers.

Smith, Rebecca A., and Susan E. Mannon. 2020. "Ethnic Entrepreneurship Without Ethnicity: Latino Entrepreneurs in Northern Utah." *Sociological Spectrum* 40 (1): 33–47.

SOM (Suns of Mercury). 2023. Community Nutrition and Education in Havana. https://www.sunsofmercury.com/havana.

SOM. 2024. Ancestral Plants in Rio de Janeiro. https://www.sunsofmercury.com/rio.

Søndergaard, Niels, Victor Thives, and Cristina Inoue. 2023. "The Shanghai Connection: Governing the Sustainability Impacts of Brazilian Agri-exports to China." In *Sustainability Challenges of Brazilian Agriculture*, edited by Niels Søndergaard, Camila Dias de Sá, and Ana Flávia Barros-Platiau, 19–44. New York: Springer.

Steel, Carolyn. 2009. *Hungry City: How Food Shapes our Lives*. London: Vintage.

Stevenson, G. W., and Richard Pirog. 2008. Values-Based Supply Chains: Strategies for Agrifood Enterprises of the Middle. In *Food and the Mid-Level Farm: Renewing an Agriculture of the Middle*, edited by Thomas A. Lyson, G. W. Stevenson, and Rick Welsh, 119–43. Cambridge: MIT Press.

Stockland. 2017. "About Stockland." http://www.stockland.com.au/about-stockland.

Stokes, Freg J. 2022. "The Hummingbird's Atlas: Mapping Guaraní Resistance in the Atlantic Rainforest during the Emergence of Capitalism, 1500–1768." PhD thesis, University of Melbourne.

Strachan, Peter. 2015. "Selandra Rise Estate to Get Community Garden." *The Herald Sun*, April 7. http://www.heraldsun.com.au/leader/south-east/selandra-rise-estate-to-get-community-garden/news-story/406575686b3dfd3e5109f8d9732ab4a1.

Strauss, Julia C., and Ariel C. Armony, eds. 2012. *From the Great Wall to the New World: China and Latin America in the 21st Century*. Cambridge: Cambridge University Press.

Strobel, Warren P., and Gordon Lubold. 2023. "Cuba to Host Secret Chinese Spy Base Focusing on U.S." *The Wall Street Journal*, June 8. https://www.wsj.com/articles/cuba-to-host-secret-chinese-spy-base-focusing-on-u-s-b2fed0e0.

Suárez Fernández, Claudia Thalía. 2024. "Se prevé mejorar la situación económica y productiva del sector azucarero." *Granma*, July 15. https://www.granma.cu/cuba/2024-07-15/se-preve-mejorar-la-situacion-economica-y-productiva-del-sector-azucarero-15-07-2024-22-07-00.

Sugiyama, Takemi, Katrien Wijndaele, Mohammad Javad Koohsari, Stephanie K. Tanamas, David W. Dunstan, and Neville Owen. 2016. "Adverse Associations of Car Time with Markers of Cardio-Metabolic Risk." *Preventive Medicine* 83: 26–30.

Sun, Ye. 2013. "Approach Brings People Back to Farms." *China Daily*, July 14. https://www.usa.chinadaily.com.cn/china/2013-07/14/content_16773035.htm.

Tang, Ming Chun. 2016. "Could U.S. Trade Threaten Sustainable Agriculture in Cuba?" *NACLA Report*, May 3. https://nacla.org/news/2016/05/02/could-us-trade-threaten-sustainable-agriculture-cuba.

Taylor, Frank F. 1989. "Cuba's Relations with the Soviet Union since October, 1962—A Retrospect." *Boletín de Estudios Latinoamericanos y del Caribe* 46: 1–109.

Tewari, Meenu, Sophie Kelmenson, Andrew Guinn, Gabriel Cumming, and Rudolph Colloredo-Mansfeld. 2018. "Mission-Driven Intermediaries as Anchors of the Middle Ground in the American Food System: Evidence from Warrenton, NC." *Culture, Agriculture, Food and Environment* 40 (2): 114–123.

Thiemann, Louis, and Max Spoor. 2019. "Beyond the 'Special Period': Land Reform, Supermarkets and the Prospects for Peasant-Driven Food Sovereignty in Post-Socialist Cuba (2008–2017)." *Canadian Journal of Development Studies* 40 (4): 546–563.

Thompson, Jennifer J., and Ashley Stinnett. 2018. "Confronting the Goldilocks Problem: Encountering 'The Middle' in Anthropological Studies of Food and Agriculture." *Culture, Agriculture, Food and Environment* 40 (2): 75–84.

Tsokhas, Kosmas. 1980. "The Political Economy of Cuban Dependence on the Soviet Union." *Theory and Society* 9 (2): 319–362.

Tsoni, Ioanna. 2016. "'They Won't Let Us Come, They Won't Let Us Stay, They Won't Let Us Leave': Liminality in the Aegean Borderscape: The Case of Irregular Migrants, Volunteers and Locals on Lesvos." *Human Geography* 9 (2): 35–46.

UEA (Urban Ecology Australia). 2019. "What is a Ecocity?" www.urbanecology.org.au/eco-cities/what-is-an-ecocity/

UN-COMTRADE. 2024. United Nations Commodity Trade Statistics Database. https://comtradeplus.un.org.

UN-DESA (United Nations Department of Economic and Social Affairs). 2018. "68% of the World Population Projected to Live in Urban Areas by 2050, says UN." *News*, May 16. http://www.un.org/development/desa/en/news/population/2018-revision-of-world-urbanization-prospects.html.

UN-DESA. 2021. Juncao Technology. https://sdgs.un.org/topics/capacity-development/juncao.

Urban, Rebecca. 2018. "Victorians' Concerns for Fast-Growing City." *The Australian*, November 5. http://www.theaustralian.com.au/news/nation/victorians-concerns-for-fastgrowing-city/news-story/af78fd4fc04e8bd7987aff3c128858b6.

Uslaner, Eric M., and Mitchell Brown. 2005. "Inequality, Trust, and Civic Engagement." *American Policy Research* 33 (6): 868–894.

Valdes, Constanza. 2022. "Brazil's Momentum as a Global Agricultural Supplier Faces Headwinds." USDA Economic Research Service, *Amber Waves*, September 27.

https://www.ers.usda.gov/amber-waves/2022/september/brazil-s-momentum-as-a-global-agricultural-supplier-faces-headwinds/.

Vannebo, Berit Irene, and Elin Birgitte Ljunggren. 2021. "Bridging or Bonding: An Organizational Framework for Studying Social Capital in Kindergartens." *International Journal of Environmental Research and Public Health* 18 (5). https://doi.org/10.3390/ijerph18052663.

Vázquez, Luis. 2020. *La territorializacion de la produccion agroalimentaria sobre bases agroecologicas en Cuba*. Habana: CELIA. http://www.agroecologynetwork.org/uploads/4/9/2/9/49299363/celia_luis_vazquez-_agroeco_cuba-covid19.pdf.

VicHealth. 2023. *Overweight and Obesity: Report of the Chief Health Officer*. Melbourne: Victorian Department of Health.

Vieira de Melo, José Evando. 2006. "Café com açucar: a formação do mercado consumidor de açucar em São Paulo e o nascimento da grande indústria açucareira paulista na segunda metade do século XIX." *Saeculum Revista de Historia* 14: 74–93.

Villapol, Nitza. 1992. *Cocina Cubana*. Havana: Editorial Científico Técnica.

Wang, Ping. 2013. "The Impacts of China's Peaceful Influence on U.S.-Mexican Relations: A Triangular Perspective." In *China and the New Triangular Relationships in the Americas: China and the Future of U.S.-Mexico Relations*, edited by Enrique Dussel Peters, Adrian H. Hearn, and Harley Shaiken, 25–35. Coral Gables, FL: University of Miami Center for Latin American Studies.

Wang, Yiwei. 2023. *"一带一路"的灯光 [Lights Along the "Belt and Road"]*. Hefei: Anhui Children's Publishing House.

Weber, Max. 1991. "The Sociology of Charismatic Authority." In *From Max Weber: Essays in Sociology*, Hans Heinrich Garth and Charles Wright Mills, 245–264. London: Routledge. Originally published in 1921.

Wemheuer, Felix. 2014. *Famine Politics in Maoist China and the Soviet Union*. New Haven: Yale University Press.

Wen, Jiabao. 2004. "Carrying Forward the Five Principles of Peaceful Coexistence in the Promotion of Peace and Development." *Chinese Journal of International Law* 3 (2): 363–368.

Wen, Tiejun. 2005. "The Relationship between China's Strategic Changes and Its Industrialization and Capitalization." In *The Chinese Model of Development*, edited by Tian Cao, 54–87. New York: Routledge.

Weissinger, Laurin B. 2020. "Building Trust and Co-designing a Study of Trust and Co-operation: Observations from a Network Study in a High-Risk, High-Security Environment." *Social Networks* 69: 136–148. https://doi.org/10.1016/j.socnet.2020.06.005.

West, Paige. 2012. *From Modern Production to Imagined Primitive: The Social World of Coffee from Papua New Guinea*. Durham, NC: Duke University Press.

Wesz, Valdemar João, Fabiano Escher, and Tomaz Mefano Fares. 2021. "Why and How Is China Reordering the Food Regime? The Brazil-China Soy-Meat Complex and COFCO's Global Strategy in the Southern Cone." *The Journal of Peasant Studies* 50 (4): 1376–1404.

While, Aidan, Andrew Jonas, and David Gibbs. 2004. "The Environment and the Entrepreneurial City: Searching for the Urban 'Sustainability Fix' in Manchester and Leeds." *International Journal of Urban and Regional Research* 28 (3): 549–569.

Wilk, Richard R. 1999. "'Real Belizean Food': Building Local Identity in the Transnational Caribbean." *American Anthropologist* 101 (2): 244–255.

Willes, Margaret. 2014. *The Gardens of the British Working Class*. New Haven and London: Yale.

Wilson, Nigel. 2014. "China's Xi Meets Fidel Castro and Praises Socialist Bond with Cuba." *International Business Times*, July 23. https://www.ibtimes.co.uk/chinas-xi-meets-fidel-castro-praises-socialist-bond-cuba-1457904.

Winter, Brian, and Caroline Stauffer. 2013. "Insight: Chinese Investors Sour on Brazil, and Projects Melt Away." *Reuters*, November 1.

Woolcock, Michael. 1998. "Social Capital and Economic Development: Toward a Theoretical Synthesis and Policy Framework." *Theory and Society* 27 (2): 151–208.

World Health Organization. 2016. "Rate of Diabetes in China 'Explosive.'" WHO Representative Office-China, news release, April 6. http://www.wpro.who.int/china/mediacentre/releases/2016/20160406/en/.

World Wildlife Fund. 2021. Sustainable Agriculture: Soy. https://www.worldwildlife.org/industries/soy.

Wright, Julia. 2009. *Sustainable Agriculture and Food Security in an Era of Oil Scarcity: Lessons from Cuba*. London: Earthscan.

Wright, Julia, and Emily Morris. 2015. "Here's What Better Relations with the US Mean for City Farms in Cuba." *The Conversation*, June 18. https://theconversation.com.

Wright, Shane, and Jennifer Duke. 2021. "Melburnians Head for the Country and Brisbane in Wake of Virus." *The Age*, August 3. https://www.theage.com.au/politics/federal/melburnians-head-for-the-country-and-brisbane-in-wake-of-virus-20210803-p58fij.html.

Xi, Jinping. 2023. "Full Text of Xi Jinping's Keynote Speech at 3rd Belt and Road Forum for Int'l Cooperation." *Xinhua*, October 18. https://www.chinadaily.com.cn/a/202310/18/WS652fa59aa31090682a5e94e1.html.

Xinhua. 2013. "China's Over-Reliance on Foreign Soybeans Worries Farmers." March 16. https://www.chinadaily.com.cn/business/2013-03/16/content_16313525.htm.

Xinhua. 2014. "China Unveils Landmark Urbanization Plan." *Xinhua*, March 17. http://www.chinadaily.com.cn/business/2014-03/17/content_17350744_2.htm.

Xinhua. 2018. "China Focus: Xi Holds Talks with Cuban President to Advance Ties." *Xinhuanet Editorial*, November 8. http://www.xinhuanet.com/english/2018-11/08/c_137592848.htm.

Xu, Yongqian. 2023. "A Particularly Dedicated Business Gentleman." *Business Weekly*, October 21. https://mp.weixin.qq.com/s/6CsZ7eokxTp0Fyynum5Zhg.

Yan, Yunxiang. 2012. "Food Safety and Social Risk in Contemporary China." *Journal of Asian Studies* 71 (3): 705–729.

Ye, Jingzhong, Baoying Ding, and Wen Wang. 2012. "A Unique Approach: Spontaneous Nested Market and Rural Development." *Chinese Rural Economy* 10: 4–12.

Zalles, Viviana, Matthew C. Hansen, Peter V. Potapov, and Selena Chavez. 2019. "Near Doubling of Brazil's Intensive Row Crop Area since 2000." *PNAS* 116 (2): 428–435.

Zhang, Li, and Gubo Qi. 2015. "On Differential Trust on Food." *Social Sciences of Beijing* 1: 36–43.

Zhang, Li, and Gubo Qi. 2018. "Bottom-up Self-Protection Responses to China's Food Safety Crisis." *Canadian Journal of Development Studies* 40 (1): 113–130.

Zhao, X. D., and S. S. Wang. 2013. "Food Trust: The Concept of Chinese Dietary and Its Transformation." *The Journal of Jiangsu Administration Institute* 24 (2): 75–80.

Zhou, Li, Sumei Pan, and Xiaoyu Dong. 2012. "From Who Will Feed China to How to Feed China: Food Properties, Model A and B of Food System and Food Sovereignty of the Development Era." *China Agricultural University Journal of Social Sciences* 29 (2): 20–33.

Zukin, Sharon. 2011. "Reconstructing the Authenticity of Place." *Theory and Society* 40 (2): 161–165.

Index

Publisher contact:
The MIT Press
Massachusetts Institute of Technology
77 Massachusetts Avenue, Cambridge, MA 02139
mitpress.mit.edu

EU Authorised Representative:
Easy Access System Europe, Mustamäe tee 50,
10621 Tallinn, Estonia
gpsr.requests@easproject.com

Printed by Integrated Books International,
United States of America